WHO Food Additives Series: 20

Toxicological evaluation of certain food additives and contaminants

Toxicological evaluation of certain food additives and contaminants

Prepared by

The 29th Meeting of the Joint FAO/WHO Expert Committee on Food Additives

Geneva, 3 - 12 June 1985

Published on behalf of

The World Health Organization

The right of the
University of Cambridge
to print and sell
all manner of books
was granted by
Henry VIII in 1534.
The University has printed
and published continuously
since 1584.

CAMBRIDGE UNIVERSITY PRESS

Cambridge

London New York New Rochelle

Melbourne Sydney

Published by the Press Syndicate of the University of Cambridge
The Pitt Building, Trumpington Street, Cambridge CB2 1RP
32 East 57th Street, New York, NY 10022, USA
10 Stamford Road, Oakleigh, Melbourne 3166, Australia

Printed in Great Britain at the University Press, Cambridge

British Library cataloguing in publication data

Joint FAO/WHO Expert Committee on Food additives,
Meeting (29th : 1985 : Geneva)
Toxicological evaluation of certain food additives and
contaminants. — (WHO Food Additives series; 20)
1. Food additives — Toxicology 2. Food contamination
3. Food poisoning
I. Title II. Series 363.1'92 RA1258

Library of Congress cataloging in publication data available

ISBN 0 521 34347 X

The preparation of this document was supported by the International Programme on
Chemical Safety (IPCS), Geneva, Switzerland.

CONTENTS

PREFACE

The monographs contained in this volume were prepared by the twenty-ninth Joint FAO/WHO Expert Committee on Food Additives (JECFA), which met in Geneva, Switzerland, 3-12 June 1985. These monographs summarize the safety data on selected food additives and contaminants reviewed by the Committee. Generally, the compounds on which monographs were prepared are those on which substantial safety data exist. The data reviewed in these monographs form the basis for acceptable daily intakes (ADIs) established by the Committee.

The twenty-ninth report of JECFA has been published by the World Health Organization as WHO Technical Report Series, No. 733. The participants in the meeting are listed in Annex 3 of the present publication and a summary of the conclusions of the Committee is included as Annex 4.

Specifications established by the twenty-ninth JECFA have been issued separately by FAO under the title *Specifications for the identity and purity of certain food additives*, FAO Food and Nutrition Paper, No. 34. These toxicology monographs should be read in conjunction with the specifications and the report.

Reports and other documents resulting from previous meetings of the Joint FAO/WHO Expert Committee on Food Additives are listed in Annex 1.

JECFA serves as a scientific advisory body to FAO, WHO, their Member States, and the Codex Alimentarius Commission, primarily through the Codex Committee on Food Additives, regarding the safety of food additives and contaminants in food. Committees accomplish this task by preparing reports of their meetings and publishing specifications and toxicological monographs, such as those contained in this volume, on substances that they have considered.

The toxicological monographs contained in this volume are based upon working papers that were prepared by temporary advisers in advance of the 1985 JECFA meeting. A special acknowledgement is given to those who prepared these working papers: Dr C.L.Galli, Professor of Experimental Toxicology, University of Milan, Milan, Italy; Dr S.I.Shibko, Associate Director of Regulatory Evaluation, Center for Food Safety and Applied Nutrition, Food and Drug Administration, Washington, DC, USA; and Dr Ronald Walker, Professor of Biochemistry, University of Surrey, Guildford, Surrey, England.

Many proprietary unpublished reports are referenced. These were voluntarily submitted to the Committee by various producers of the food additives under review and in many cases these reports represent the only safety data available on these substances. The temporary advisers based the working papers they developed on all the data that were submitted, and all these studies were available to the Committee when it made its evaluations.

From 1972 to 1975 the toxicology monographs prepared by Joint FAO/WHO Expert Committees on Food Additives were published by WHO in the WHO Food Additives Series; after 1975 this series became available only in the form of unpublished WHO documents provided on request by the Organization. Henceforth their publication by Cambridge University Press should ensure that these monographs are more widely known and available.

The preparation and editing of the monographs included in this volume have been made possible through the technical and financial contributions of the Participating Institutions of the International Programme on Chemical Safety (IPCS), which support the activities of JECFA. IPCS is a joint venture of the United Nations Environment Programme, the International Labour Organisation, and the World Health Organization, which is the executing agency. One of the main objectives of IPCS is to carry out and disseminate evaluations of the effects of chemicals on human health and the quality of the environment.

The designations employed and the presentation of the material in this publication do not imply the expression of any opinion whatsoever on the part of the organizations participating in IPCS concerning the legal status of any country, territory, city, or area or its authorities, or concerning the delimitation of its frontiers or boundaries. The mention of specific companies or of certain manufacturers' products does not imply that they are endorsed or recommended by those organizations in preference to those of a similar nature that are not so mentioned.

Any comments or new information on the biological or toxicological data on the compounds reported in this document should be addressed to: Joint WHO Secretary of the Joint FAO/WHO Expert Committee on Food Additives, International Programme on Chemical Safety, World Health Organization, Avenue Appia, 1211 Geneva 27, Switzerland.

ENZYME PREPARATIONS AND ENZYME IMMOBILIZING AGENTS

CARBOHYDRASE (α-AMYLASE) FROM BACILLUS LICHENFORMIS

EXPLANATION

Carbohydrase is an enzyme that catalyzes the hydrolysis of α-1,4-glycosidic linkages of starch. The enzyme preparation that is derived from B. lichenformis is added directly to the food to be processed and then it is removed from the final product by filtration. This preparation has not been previously evaluated by the Joint FAO/WHO Expert Committee on Food Additives.

BIOLOGICAL DATA

Biochemical aspects

No information available.

Toxicological studies

Special studies on genetic toxicity

Groups of 20 CD-1 male mice fed diets containing 0, 1.0, 2.0, or 4.0% of the carbohydrase preparation were used in a dominant-lethal study. The animals were fed the test compound for 5 days. These males were then mated 1-to-1 at weekly intervals with 5 different batches of 20 females each. At 14 days, after evidence of copulation, females were sacrificed and the uterine contents examined for implantations, viable embryos, and early and late embryonic deaths. Some weight loss occurred in both the high- and mid-dose males. One total-litter loss occurred in each of the high- and mid-dose groups at the second pairing and in the low-dose group at the third pairing, but the incidence of these losses was too low to be considered treatment-related and there were no other compound-related effects (Palmer & Lowell, 1973a).

A dominant-lethal study was carried out using groups of 20 male CD rats given a diet containing 0, 1, 2, or 4% of the carbohydrase preparation for 5 days. The males were then mated on a 1-to-1 basis for 7 days with untreated CD females. A new batch of females were mated with treated males every 7 days for 6 consecutive weeks. Pregnant females were sacrificed on about day 14 of pregnancy and ovaries and uteri were examined for corpora lutea, implantations, viable embryos, and early and late embryonic deaths. During the 5-day treatment period weight loss occurred in the high-dose males, and body-weight gains were reduced at the mid-dose level. There was no significant effect of treatment on mating performance, pregnancy rate, pre- or post-implantation loss, or overall viable litter size. Although 2 females in the high-dose group suffered total-litter losses, the overall incidence did not suggest a treatment-related effect (Palmer & Lowell, 1973b).

Special study on teratology

Groups of 20 mated female CD rats were fed diets containing 0, 1, 2, or 4% of the carbohydrase preparation from days 6 through 15 of pregnancy. The animals were sacrificed on day 20 of pregnancy and the uterine contents examined for the number of corpora lutea, number of viable foetuses, number of resorption sites, litter weight, and foetal abnormalities. One-third of the foetuses were examined by the Wilson technique for visceral abnormalities and two-thirds were processed for examination of skeletal abnormalities. During the first 4 days of dietary administration of the test compound there was reduced dietary consumption and reduced weight gain at all dose-levels (including weight loss at the high dose). No compound-related effects on foetal or embryological development were reported, although there was a non-significant increase in foetal abnormalities at the high dose (Palmer & Lowell, 1972).

Acute toxicity

Species	Route	LD$_{50}$ (mg/kg b.w.)	Reference
Mouse	oral (gavage)	20,000	Novo, 1973a
Rat (male)	oral (gavage)	20,500	Novo, 1973b
(female)		16,500	

Short-term studies

Rats

Groups of 5 male and 5 female Wistar rats were fed diets containing 0, 0.5, or 2.5% of the carbohydrase preparation in the diet for 4 weeks. Except for a small but statistically-significant increase in absolute and relative kidney weights in the high-dose males, and some fluctuation in weight gain during the initial part of the study, there were no compound-related changes. Gross pathology, clinical chemistry, and feed efficiency were comparable between groups (Novo, 1972).

Groups of 15 male and 15 female CFY-strain rats were fed diets containing 0, 1, 2, and 4% of the carbohydrase preparation in the diet for 13 weeks. Thinning of the hair, mainly on the scapular region, was noted in 7 of the high-dose female rats from week 9 onward. There was reduced feed intake in mid-dose females and high-dose males and females, and decreased weight gain in high-dose animals of both sexes. Organ to body-weight ratios for several organs from the high-dose animals differed significantly from control values; however, many of these differences were likely to have arisen because of reduced body-weight gain. When compared to brain weights, only reduced liver weights in high-dose males were found to be significantly differently from controls. Increased adrenal weights in high-dose females were considered to be within the normal range of biological variability. Enlargement of the caecum was noted in mid-dose males and in both sexes at the high dose. No compound-related changes were reported with regard to survival, urinalysis, haematology, clinical chemistry, or microscopic pathology (Rivett <u>et al</u>., 1973a).

Dogs

Groups of 3 male and 3 female Beagle dogs were given diets containing 0, 1, 2, or 4% carbohydrase preparation for 13 weeks. Reduced mean body-weight gain was observed in high-dose animals of both sexes. Food and water consumption were reduced in high-dose animals of both sexes and in mid-dose females. No compound-related changes were observed with respect to haematology (1 high-dose female had platelet counts greater than the normal range), clinical chemistry, urinalysis, or gross and microscopic pathology. Differences between control and high-dose animals with respect to organ-weight ratios were ascribed to reduced growth of the high-dose animals (Rivett et al., 1973b).

Long-term studies

No information available.

Observations in man

No information available.

Comments

The carbohydrase preparation showed no significant toxicological effects in short-term feeding studies in rats at levels of up to 4% of the diet (40 mg/kg of feed) or in dogs at levels of up to 2% (20 mg/kg of feed). No teratogenic effects were noted in a study in rats. The preparation was also inactive in dominant-lethal tests in rats and mice.

EVALUATION

Level causing no toxicological effect

The no-effect level in a short-term study in dogs was 2% of the diet, equal to 450 mg/kg b.w.

Estimate of acceptable daily intake for man

ADI "not specified".

REFERENCES

Novo (1972). Four week oral toxicicity study of Novo alkaline amylase in rats. Unpublished study by Novo Industria A/S. Submitted to WHO by Novo Industria A/S.

Novo (1973a). Acute toxicity of Novo amylase to mice. Unpublished study by Novo Industria A/S. Submitted to WHO by Novo Industria A/S.

Novo (1973b). Acute toxicity of Novo alkaline amylase to rats. Unpublished study by Novo Industria A/S. Submitted to WHO by Novo Industria A/S.

Palmer, A.K. & Lovell, M.R. (1972). Effect of NAA on pregnancy of the rat. Unpublished report of the Huntingdon Research Centre. Submitted to WHO by Novo Industria A/S.

Palmer, A.K. & Lovell, M.R. (1973a). Dominant lethal assay of NAA in the male mouse. Unpublished report of the Huntingdon Research Centre. Submitted to WHO by Novo Industria A/S.

Palmer, A.K. & Lovell, M.R. (1973b). Dominant lethal assay of NAA in the male rat. Unpublished report of the Huntingdon Research Centre. Submitted to WHO by Novo Industria A/S.

Rivett, K.F., Bhutt, A., Street, A.E., Heywood, R., & Newman, A.J. (1973a). Novo alkaline amylase dietary study in rats. Unpublished report of the Huntingdon Research Centre. Submitted to WHO by Novo Industria A/S.

Rivett, K.F., Sortwell, R.J., Newman, A.J., & Street, A.E. (1973b). Novo NAA toxicity studies in beagle dogs. Unpublished report of the Huntingdon Research Centre. Submitted to WHO by Novo Industria A/S.

GLUCOSE ISOMERASE (IMMOBILIZED) FROM
ACTINOPLANES MISSOURIENSIS

EXPLANATION

This enzyme preparation has not been previously evaluated by the Joint FAO/WHO Expert Committee on Food Additives.

BIOLOGICAL DATA
Biochemical aspects

No information available.

Toxicological studies
Special study on reproduction

Rats

Groups of 20 male and 20 female Sprague-Dawley rats were fed diets containing 0 or 1% whole, non-viable Actinoplanes missouriensis in the diet. After 60 days on test, males and females were mated on a 1-to-1 basis. The males were sacrificed after 90 days on test for organ-weight analysis and gross and microscopic pathology evaluation. The females were allowed to litter and nurse their young until weaning at 21 days. Body weights tended to be lower in male rats given the test compound but did not differ significantly from controls. Right gonad weights tended to be slightly higher in treated males and females. No compound-related changes were observed in haematology, urinalysis, clinical chemistry, or gross or microscopic pathology, nor were there any effects on reproductive performance of the offspring (Tisdel & Harris, 1974a).

Acute toxicity

Species	Route	LD$_{50}$ (mg/kg b.w.)	Reference
Rat (male)	Oral (dietary)	40,000	Regel, 1973
Mouse (male)	i.v.	1,250	Regel, 1973
Mouse (male)	i.v.	875–1,250	Regel, 1973
Rabbit	s.c.	1,250	Regel, 1973

Short-term study
Dogs

Groups of 2 male and 2 female beagle dogs were fed for 90 days diets containing 0 or 1% whole, non-viable cells of __Actinoplanes missouriensis__. Body-weight gains and food consumption were less for the test dogs than controls, but remained in the normal range for dogs of their age during the course of the study. Results of haematology and urinalysis were normal for all animals and no compound-related effects on organ weights or gross and microscopic pathology were reported (Tisdel & Harris, 1974b).

Long-term studies
No information available.

Observations in man
No information available.

Comments
A well-conducted short-term study in rats, which included a 1-generation reproduction study, showed no significant toxicological effects. A short-term study in dogs provides additional information on the lack of toxicity of the preparation.

The studies on the non-immobilized enzyme were considered by the Committee to be appropriate for evaluating the immobilized form because the use of gelatin as an entrapping agent does not present a toxicological problem. The release of free glutaraldehyde from the enzyme preparation is controlled by the specifications for the preparation established by the Committee.

EVALUATION
Level causing no toxicological effect
Rat: 1% (10,000 ppm) in the diet, equivalent to 1,000
        mg/kg b.w./day.

Estimate of acceptable daily intake for man
Acceptable for use in food processing when used as a component in an immobilized system.

REFERENCES

Regel, L. (1973). Untitled report on acute toxicity studies. Unpublished study of the Warf Institute, Inc., Madison, WI, USA. Submitted to the World Health Organization by Gist-Brocades.

Tisdel, M. & Harris, D. (1974a). 90-Day subacute and reproduction study - Rat. Unpublished study of the Warf Institute, Inc., Madison, WI, USA. Submitted to the World Health Organization by Gist-Brocades.

Tisdel, M. & Harris, D. (1974b). 90-Day subacute and reproduction study - Dog. Unpublished study of the Warf Institute, Inc., Madison, WI, USA. Submitted to the World Health Organization by Gist-Brocades.

GLUCOSE ISOMERASE (IMMOBILIZED) FROM BACILLUS COAGULANS

EXPLANATION

This enzyme preparation has not been previously evaluated by the Joint FAO/WHO Expert Committee on Food Additives.

BIOLOGICAL DATA

Biochemical aspects

No studies available.

Toxicological studies

Special study on mutagenicity

Rats

A dominant lethal study was carried out using Sprague-Dawley rats. Groups of 12 or 15 males were dosed twice daily for 5 consecutive days by gavage with immobilized glucose isomerase in an aqueous 0.5% tragacanth gum suspension such that they received a total daily dose of 0, 3000, or 9000 mg/kg b.w. of the enzyme. A positive control group of 11 males received 3 mg/kg b.w. of thiotepa per day by i.p. injection. On the sixth day following treatment, each male was housed with two females. Every 7 days, 2 new females were paired with every male, and the procedure continued for a total of 10 weeks. Half of the females were sacrificed on day 5 of gestation, and the other half on day 21. A slight reduction in testicular weights was observed in high-dose males and a significant increase in pre-implantation losses occurred in low-dose, but not high-dose females in the 10-week mating. No compound-related effects were observed with respect to weight gain in males, fertilization index, development of fertilized ova, or post-implantation loss (Tesh, 1976).

Special study on reproduction
Rats

A 3-generation reproduction study was carried out using groups of 15 male and 30 female Sprague-Dawley rats given 0, 0.5, 1.0, or 5.0% immobilized glucose isomerase in the diet. These dietary concentrations were given for 70 days prior to and through mating (on a 1 male to 2 female basis) and throughout 2 successive pregnancies in each of 3 generations. The litters arising from the first pregnancy were sacrificed after 14 days of lactation and the second litter of each generation stayed with the dam until weaning, when 15 males and 30 females were selected at random to produce the succeeding generation. After each round of mating had been completed, 5 pregnant females from each group were sacrificed at day 14 of gestation for examination of uterine contents. A total of 20 males and 20 females from the F_{3b} litter were sacrificed after weaning for gross and microscopic pathological examination. No compound-related effects were reported on body-weight gain of parental animals of either sex, food consumption, litter weights, mating performance, reproductive indices, foetal development, or gross or microscopic pathology. There was an increased incidence of females with irregular estrous cycles and increased pre-coital intervals in the mid- and high-dose groups. However, there was no effect on reproductive performance in these groups. There was a slight increase in gonadal weights in the mid- and high-dose F_{3b} males, but no gross or microscopic changes related to treatment were found (Tesh & Smith, 1976).

Special studies on teratogenicity
Rats

Groups of 20 pregnant female CD rats were fed diets containing 0, 0.5, 1.0, or 5.0% immobilized glucose isomerase in the diet from days 6 through 15 of gestation. The animals were sacrificed on day 21 of pregnancy; one-third of the foetuses were processed for examination of soft tissue anomalies and the remaining foetuses were processed for examination of skeletal abnormalities. There was an increase in pre-implantation loss in the high-dose group; however, no effect was reported on the number of viable foetuses, and the pre-implantation effect seemed to be due to an increase in corpora lutea in the high-dose rats (Tesh et al., 1975).

Rabbits

Groups of 20 pregnant New Zealand-strain rabbits were dosed by gavage with 0, 250, 500, or 750 mg/kg b.w. immobilized glucose isomerase (in an aqueous 0.5% tragacanth-gum vehicle) from days 6 through 18 of pregnancy. The animals were sacrificed on day 29 of pregnancy. Foetuses were examined internally and externally and then processed for skeletal examination. No compound-related effects were observed on maternal weight-gain, pre- or post-implantation loss, litter size, foetal weight or foetal anomalies (Tesh & Toseland, 1975).

Acute toxicity

Species	Route	LD_{50} (mg/kg b.w.)	Reference
Mouse	oral (gavage)	6,000	Novo, 1974a
Rat	oral (gavage)	6,000	Novo, 1974b
Dog	oral (gavage)	5,000	Novo, 1975

Short-term studies

Rats

Groups of five female Wistar rats were fed diets containing 0, 1, or 10% immobilized glucose isomerase for 4 weeks. Weight gain was slightly greater in the treated animals and feed intake was significantly higher in males and females in both treatment groups. Feed conversion efficiency was slightly reduced in dosed females during the latter half of the study. Haemoglobin concentrations were lower in treated as compared to control animals; however, this was attributed to abnormally high values in the control group. No treatment-related changes were reported with respect to clinical chemistry, absolute and relative organ weights, or gross and microscopic pathology (Novo, 1974c).

Groups of 15 male and 15 female Sprague-Dawley rats were given diets containing 0, 0.5, 1.0, or 5.0% immobilized glucose isomerase for 13 weeks. No significant compound-related effects were reported with regard to body-weight gain, feed consumption, haematology, blood chemistry, urinalysis, opthalmoscopy, absolute or relative organ-weights, or gross pathology. There was an increase in incidence and size of focal mineralization in the kidneys of treated females as opposed to controls. No other compound-related histological changes were observed.

Another study was performed to investigate the cause of the renal mineralization. Groups of 15 female Sprague-Dawley rats were fed one of two different commercial laboratory diets containing 0, 0.1, 1.0, or 5.0% immobilized glucose isomerase. Body-weight gain, urinalysis, organ weights, and gross pathology were not significantly affected by treatment. There was no effect of treatment on animals fed one of the laboratory diets. In the animals fed the other diet, the incidence of focal mineralization was greater in the high-dose group than in the controls, but the other dose groups did not show a treatment-related effect (Wheldon & Ben-Dyke, 1975).

Dogs

Groups of 3 male and 3 female beagle dogs were fed diets containing 0, 0.5, 1.0, or 5.0% immobilized glucose isomerase in the diet for 13 weeks. No compound-related effects were observed with respect to body-weight gain, ophthalmoscopy, haematology, blood and urine chemistry, organ weights, or gross or microscopic pathology (Wheldon & Ben-Dyke, 1975).

Long-term studies
No information available.

Observations in man
No information available.

Comments
The focal mineralization of the kidney of the rat in one study was restricted to females receiving 5% of the enzyme prepara-

tion. Females fed a different laboratory diet containing 5% of the enzyme preparation did not exhibit this lesion. No other significant compound-related effects were observed in either rat study. No adverse effects were observed in a short-term feeding study in dogs, teratology studies in rats and rabbits, or reproduction and dominant lethal studies in rats.

EVALUATION

Level causing no toxicological effect

Rat: 5% (50,000 ppm) in the diet, equivalent to 5000 mg/kg
       b.w./day.

Estimate of acceptable daily intake for man

Acceptable for use in food processing when used as a component in an immobilized system.

REFERENCES

Novo (1974a).   Acute oral toxicity of SP113 to mice. Unpublished report of Novo Industria A/S. Submitted to the World Health Organization by Novo Industria A/S.

Novo (1974b).   Acute oral toxicity of SP113 to rats. Unpublished report of Novo Industria A/S. Submitted to the World Health Organization by Novo Industria A/S.

Novo (1974c).   Four week oral toxicity study of glucose isomerase (SP113) in rats. Unpublished report of Novo Industria A/S. Submitted to the World Health Organization by Novo Industria A/S.

Novo (1975).   Acute toxicity of glucose isomerase-SP113 given once perorally to Beagle dogs. Unpublished report of Novo Industria A/S. Submitted to the World Health Organization by Novo Industria A/S.

Tesh, J.M. (1976).   SP113: Dominant lethal study in rats. Unpublished report of Life Science Research. Submitted to the World Health Organization by Novo Industria A/S.

Tesh, J.M.   &   Smith, C. (1976).   SP113:   Multigeneration study in rats. Unpublished report of Life Science Research. Submitted to the World Health Organization by Novo Industria A/S.

Tesh, J.M.   & Toseland, A.E. (1975).   SP113: Effects upon pregnancy in the rabbit. Unpublished report of Life Science Research. Submitted to the World Health Organization by Novo Industria A/S.

Tesh, J.M., Toseland, A.E., & Pinson, C.D. (1975). SP113: Effects upon pregnancy in the rat. Unpublished report of Life Science Research. Submitted to the World Health Organization by Novo Industria A/S.

Wheldon, G.H., Ben-Dyke, R., Perry, M.C., & Newman, A.J. (1975). SP113: Toxicity in continuous dietary administration to rats for 13 weeks. Unpublished report of Life Science Research. Submitted to the World Health Organization by Novo Industria A/S.

Wheldon, G.H., & Ben-Dyke, R. (1975).   SP113: Toxicity in dietary administration to dogs over 13 weeks. Unpublished report of Life Science Research. Submitted to the World Health Organization by Novo Industria A/S.

GLUCOSE ISOMERASE (IMMOBILIZED) FROM <u>STREPTOMYCES OLIVACEUS</u>

EXPLANATION

This enzyme preparation has not previously been evaluated by the Joint FAO/WHO Expert Committee on Food Additives.

BIOLOGICAL DATA
Biochemical aspects

No information available.

Toxicological studies
Special study on reproduction

Rats

Groups of 10 male and 10 female Charles-River CD rats were fed diets containing 0, 1.5, 3.0, or 6.0% immobilized glucose isomerase for 100 days, at which time they were mated on a one-to-one basis. Groups of 15 males and 15 females (F_1 generation) that resulted from the mating were fed the same diets for 90 days; the animals were then sacrificed for gross and microscopic pathology studies. No compound-related effects were reported with respect to fertility indices, body weights, food consumption, ophthalmoscopy, haematology, clinical chemistry, urinalysis, or gross or microscopic pathology (Geil et al, 1975).

Acute toxicity

Species	Material tested	Route	LD$_{50}$ (mg/kg b.w.)	Reference
Rat	<u>S. olivaceus</u> cells	oral (gavage)	2500	Hartnagel, 1970
Rat	<u>S. olivaceus</u> fermenter beer	oral (gavage)	50 ml/kg	Porter & Hartnagel, 1974
Rat	non-immobilized glucose isomerase	oral (gavage)	3330	Hartnagel, 1971
Rat	immobilized glucose isomerase	oral (gavage)	3330	Hartnagel, 1971

Short-term studies
Rats

Groups of 15 male and 15 female Charles-River CD rats were
fed diets containing 0, 1.5, 3.0, or 6.0% immobilized glucose isomerase
for 90 days. Haematology, clinical chemistry, and urinalysis
measurements were conducted on 5 rats/sex/dose at 90 days, and on 5
male and 5 female control and high-dose animals at 30 and 60 days.
Gross pathology studies were conducted on all animals and microscopic
pathology was conducted on 5 males and 5 females from the high-dose
groups and controls. Total leucocyte counts were elevated in the high-
and mid-dose animals at 90 days. There were no reported
compound-related changes with respect to body-weight gain, food
consumption, ophthalmoscopy, clinical chemistry, urinalysis, organ
weights, or gross or microscopic pathology (Benson <u>et al</u>., 1975).

Dogs

Groups of 4 male and 4 female beagle dogs were fed diets
containing 0, 1.5, 3.0, or 6.0% immobilized glucose isomerase in the
diet for 90 days. One female control dog died during the study; the
likely cause of death was pneumonia. Other dogs on test also developed
acute pneumonia, possibly related to rooting and sniffing in the
powdered diet. The pathology report indicated that acute exudative
(lung) lesions noted in control and all treatment groups likely

resulted from secondary bacterial infection in areas of the lung previously damaged by lung worms (Cookson et al., 1975).

Because of the lung pathology noted above in the first dog study, a new study was undertaken. Groups of 4 male and 4 female beagle dogs were fed diets containing 0, 1.5, 3.0, or 6.0% immobilized glucose isomerase for 90 days. No compound-related changes occurred with respect to body-weight gain, food consumption, ophthalmoscopy, haematology, clinical chemistry, urinalysis, organ weights, or gross or microscopic pathology. Lung lesions occurred at a high incidence in all dose groups and the controls and may have been due to infection with parasitic larvae in the lungs (Harris & Teeter, 1976).

Long-term studies

No information available.

Observation in man

No information available.

Comments

No compound-related adverse effects were reported when dietary levels of up to 6% of immobilized glucose isomerase were fed to rats in reproduction and subchronic toxicity studies and to dogs in subchronic toxicity studies.

EVALUATION

Level causing no toxicological effect

Rat: 6% (60,000 ppm) in the diet, equivalent to 6000 mg/kg
     b.w./day.

Estimate of Acceptable Daily Intake for Man

Acceptable for use in food processing when used as a component in an immobilized system.

REFERENCES

Benson, B.W., Geil, R.G., Keller, W.F., & Blanchard, G.L. (1975). Subchronic studies of the toxicity of glucose isomerase enzyme: II. Ninety day feeding study in rats. Unpublished study of the International Research and Development Corporation. Submitted to the World Health Organization by Miles Laboratories, Inc.

Cookson, K.M., Geil, R.G., & Keller, W.F. (1975). Subchronic studies of the toxicity of glucose isomerase enzyme: I. Ninety day feeding study in dogs. Unpublished study of the International Research and Development Corporation. Submitted to the World Health Organization by Miles Laboratories, Inc.

Geil, R.G., Benson, B.W. Harris, S.B., & Keller, W.F. (1975). Subchronic studies of the toxicity of glucose isomerase enzyme: III. Two generation feeding study in rats. Unpublished study of the International Research and Development Corporation. Submitted to the World Health Organization by Miles Laboratories, Inc.

Harris, D.L. & Teeter, C.L. (1976). Subchronic studies of the toxicity of glucose isomerase enzyme: IV. Ninety day feeding study in dogs. Unpublished study of the WARF Institute Inc. Submitted to the World Health Organization by Miles Laboratories, Inc.

Hartnagel, R.E. (1970). The acute oral toxicity of a strain of glucose isomerase-producing S. olivaceus cells. Unpublished study of Miles Laboratories, Inc. Submitted to the World Health Organization by Miles Laboratories, Inc.

Hartnagel, R.E. (1971). The acute oral toxicity of fixed and non-fixed glucose isomerase and isomerized syrup. Unpublished study of Miles Laboratories Inc. Submitted to the World Health Organization by Miles Laboratories, Inc.

Porter, M.C. & Hartnagel, R.E. (1974). Study of the acute oral toxicity of glucose isomerase fermentor beer in the rat. Unpublished study of Miles Laboratories, Inc. Submitted to the World Health Organization by Miles Laboratories, Inc.

GLUCOSE ISOMERASE (IMMOBILIZED)
FROM STREPTOMYCES OLIVOCHROMOGENES

EXPLANATION
This substance has not been previously evaluated by the Joint FAO/WHO Expert Committee on Food Additives.

BIOLOGICAL DATA
Biochemical aspects
No information available.

Toxicological studies
Acute toxicity
No information available.

Short-term studies
Rats

Groups of 15 male and 15 female Charles-River albino rats were fed 0, 1, 3, or 10% of a preparation of whole S. olivochromogenes cells in the diet for 90 days. Harvesting of the cells for the feeding studies was performed by the addition of 36 g/litre of Perlite to the fermentation media followed by filtration and washing. (A semipurified enzyme preparation from lysed cells - no Perlite added - is used in the actual commercial production of high fructose corn syrup.) An additional group of 15 males and 15 females were fed a diet containing 7.5% Perlite for 90 days. Clinical chemistry, haematology, urinalysis, and microscopic pathology were conducted on 10 animals/sex from the control, Perlite, and high-dose groups.

There was a small decrease in liver to body-weight ratio in all treatment groups and in the Perlite groups. The decreases were not dose-related. Spleen to body-weight ratios were decreased in mid- and high-dose females, but the magnitude of the effect was small. There was an increase in ovary to body-weight ratio in low- and high-dose females, but there was an inverse dose relationship and the magnitude of the effect was small. No treatment-related changes were reported with respect to body-weight gain, food consumption, haematology, clinical chemistry, urinalysis, or gross or microscopic pathology (Smith, 1972).

Dogs

Groups of 4 male and 4 female beagle dogs were fed diets containing 0, 1, 3, or 10% of a S. olivochromogenes whole-cell preparation similar to that described above in the rat study. An additional group of 4 males and 4 females were fed a diet containing 7.5% Perlite. A significant reduction in body-weight gain was observed in the male high-dose group. No statistically-significant effects of treatment on food consumption were reported, but there appeared to be a dose-related trend toward lower food consumption in females. An increased brain to body-weight ratio in high-dose and diluent-control males was probably due to the reduced body-weight gain in these groups. No significant treatment-related changes were reported with respect to haematology, clinical chemistry, urinalysis, or gross or microscopic pathology (Burtner, 1972).

Comments

Short-term feeding studies in rats and dogs showed no significant toxicological effects. Although the preparation fed in these studies differs from that used in commercial preparations, the high levels of whole cells fed (up to 10% of the diet) plus the use of a Perlite control provide adequate information for evaluating the safety of the preparation derived from lysed cells.

EVALUATION

Level causing no toxicological effect

Rat: 10% (100,000 ppm) in the diet, equivalent to
      10,000 mg/kg b.w./day.

Estimate of acceptable daily intake for man

Acceptable for use in food processing when used as a
component in an immobilized system.

REFERENCES

Burtner, B.R. (1972). Ninety-day subacute oral toxicity study with
        XSO-1228 in Beagle dogs. Unpublished report of Industrial
        BIO-TEST Laboratories, Inc. Submitted to the World Health
        Organization by CPC International, Inc. (validated study).
Smith, P.S. (1972). Ninety-day subacute oral toxicity study with
        XSO-1228 in albino rats. Unpublished report of Industrial
        BIO-TEST Laboratories, Inc. Submitted to the World Health
        Organization by CPC International, Inc. (validated study).

GLUCOSE ISOMERASE (non-immobilized and immobilized)
from STREPTOMYCES RUBIGINOSIS

EXPLANATION

These enzymes preparations have not been previously reviewed by the Joint FAO/WHO Expert Committee on Food Additives.

BIOLOGICAL DATA

Biochemical aspects

No information available.

Toxicological studies

Special studies on reproduction

Rats

Groups of 10-12 male Wistar-derived rats were fed diets for 60 days containing 0, 0.4, or 1.6% of an enzyme preparation (non-immobilized), in use prior to 1973 (designated HFE), or 0.125 or 1.25% of an enzyme preparation immobilized to DEAE-cellulose (designated DCI), which has been used since 1973. Females kept on the same diets for 14 days were mated on a one-to-one basis with the dosed males. After one week the first set of females were replaced with a new set and mated with the males. One set of females was allowed to cast their litters normally and nurse the young until weaning at 21 days. The second set of females was sacrificed on day 13 of pregnancy and the uterine contents examined. The fertility was slightly depressed in all dosed groups as compared to the control values and the response was characterised as not unusual in animals fed semi-purified diets. Resorptions were increased as compared to controls in the high-dose HFE and DCI groups, but the magnitude of the effect was not regarded as significant. There were no compound-related effects on gestation, viability, or lactation indices, nor on corpora lutea, total implants, or live fetuses (Cox et al., 1973).

Groups of 24-29 pregnant Wistar rats were fed diets which provided 0, 200, or 800 mg/kg b.w./day of HFE, or 62.5 or 625 mg/kg b.w./day of DCI. The test compounds were fed beginning at day 14 of pregnancy and continued through parturition and lactation until weaning at day 21. No compound-related changes were observed on fertility, gestation, viability, or lactation indices (Cox et al., 1973).

Special studies on teratology
Rats
Groups of 22 pregnant female Wistar-derived rats were given doses of 0, 200, or 800 mg/kg b.w. of HFE, or doses of 62.5 or 625 mg/kg b.w. of DCI by gavage daily from days 6 through 15 of pregnancy. The animals were sacrificed and their uterine contents examined 1 day prior to the full term of pregnancy. One-third of the foetuses were subjected to visceral examination and two-thirds were processed for skeletal examination. No compound-related effects were observed on soft tissue or skeletal abnormalities, nor on number of implant sites, resorption sites, or live and dead foetuses (Cox et al., 1973).

Rabbits
Groups of 13-17 female Dutch-belted rabbits were artificially inseminated and dosed daily with 0, 200, or 800 mg/kg b.w. HFE or with 62.5 or 625 mg/kg b.w. DCI by gavage on days 6-18 of pregnancy. On day 28 the animals were subjected to Ceasarean section and the uterine contents examined. The pups were placed in an incubator for 24 hours for evaluation of neonatal survival, then examined for visceral abnormalities and processed for skeletal examination. There was a small increase in number of resorptions and/or foetal deaths in the high-dose HFE groups. There were no compound-related effects on corpora lutea, implant sites, live foetuses, or soft tissue or skeletal anomalies (Cox, et al., 1973).

Acute toxicity

No information available.

Short-term studies

Rats

Groups of 20 weanling rats/sex/dose were fed diets containing 0, 3, 6, or 12% of an enzyme "cake" of a glucose isomerase preparation for 90 days. Pathology studies were conducted on 10 animals of each sex in the high-dose and control groups. No adverse effects were found with respect to body-weight gain, clinical chemistry, haematology, relative and absolute organ weights, or gross and microscopic pathology (Oser, 1969).

A 33-week feeding study was carried out in Wistar rats. Groups of 30 males and 30 females were fed diets containing 0.4 or 4.0% HFE or 0.125 or 1.25% DCI. The animals in the high-dose HFE group showed a marked decrease in food intake and began losing weight by the ninth week. Four animals/sex in the high-dose HFE group were sacrificed and necropsied at 10 weeks. The remaining animals in the high-dose group were removed from the 4% diet and separated into 2 equal groups. One group received a basal (control) diet and one received 2% of the HFE preparation. An immediate resumption of food intake and growth ensued, and after a 2-week period all of the original animals on the high-dose diet were recombined and put on the 2% diet. An interim sacrifice of 5 animals/sex/group was conducted at week 16 for pathological examination. Laboratory analyses for haematology, clinical chemistry, and urinalysis were conducted on 10 animals/sex/ group at 6, 12, and 26 weeks. The results of these analyses indicated no compound-related effects. Gross and microscopic pathology studies and studies on organ weights at 16 and 33 weeks showed no compound-related effects of either enzyme preparation.

An additional study was conducted to determine if palatability was the cause of the weight loss exhibited in the high-dose (4%) HFE animals. Groups of 10 rats/sex were given 0 or 2000 mg/kg b.w. of the HFE preparation daily by gavage in a water vehicle (both groups received a control diet). Decreased weight gain was observed in the treated rats, more markedly in the males. However, the magnitude of the effect was much less than seen in the dietary study, indicating that palatability was likely a factor in that study (Cox et al., 1973).

Dogs
Groups of 5 dogs/sex were given diets containing 0.4 or 4.0% HFE, or 0.125 or 1.25% DCI, for 30 weeks. An interim sacrifice of 1 animal/group/sex was carried out at 11 weeks in the high-dose group and 14 weeks in the remaining groups. There were palatability problems in dogs at the 4% level of the HFE preparation. After 2 weeks, 0.1% dried garlic was added to their diet in an attempt to mask the odor of HFE. At 12 weeks, because of reduced-weight gain and palatability difficulties, the high-dose HFE diet was reduced to 2.0% Other than reduced-weight gain in the high-dose HFE dogs there were no compound-related effects on weight gain, haematology, clinical chemistry, urinalysis, organ weights, or gross or microscopic pathology (Cox et al., 1973).

Long-term studies
No information available.

Observations in man
No information available.

Comments
Decreased weight gain, due in part at least to reduced dietary palatability, was seen when high doses of the non-immobilized (HFE) form of the enzyme was fed to rats and dogs. A suggestion of a possible foetotoxic response was also seen when high doses of the enzyme were given to rats and rabbits. No gross or microscopic pathology was associated with this form of the enzyme.

The immobilized form of the enzyme showed no adverse effects in short-term feeding studies in rats and dogs, teratology studies in rats, dogs, and rabbits, or in reproduction studies in rats except for a small increase in resorptions, which was of uncertain significance, in animals given high doses of the substance.

EVALUATION

Level causing no toxicological effect

Non-immobilized enzyme

Rats: 2% (20,000 ppm) in the diet, equivalent to 2000 mg/kg
b.w./day.

Immobilized enzyme

Rats: 1.25% (12,500 ppm) in the diet, equivalent to
1250 mg/kg b.w./day.

Estimate of acceptable daily intake for man

Non-immobilized enzyme

No ADI allocated, because no information was available on
the food use of the non-immobilized enzyme.

Immobilized enzyme

Acceptable for use in food processing when used as a
component in an immobilized system.

REFERENCES

Cox, G.E., Bailey, D.E., & Morareidge, K. (1973).   Sub-acute feeding
studies in rats and dogs with glucose isomerase enzyme
preparations. Unpublished report of Food and Drug Research
Laboratories, Inc., Waverly, NY, USA. Submitted to WHO by
Nabisco Brands, Inc.

Oser, B.L. (1969).   Subacute (90 day) feeding studies with enzyme cake
and isomerase 100 (converted starch). Unpublished report of
Food and Drug Research Laboratories, Inc., Waverly, NY,
USA. Submitted to WHO by Nabisco Brands, Inc.

POLYETHYLEMINE AND ETHYLENIMINE

EXPLANATION

Polyethylenimine is an immobilizing agent used in the production of enzyme preparations for food processing. The substance is also used in food packaging materials. Polyethylenime is produced by the acid-catalyzed homopolymerization of ethylenimine. The polymerized material is cross-linked with ethylene dichloride to give the 40,000 to 60,000 molecular-weight substance utilized in enzyme immobilization.

POLYETHYLEMINE

EXPLANATION

This substance has not been previously evaluated by the Joint FAO/WHO Expert Committee on Food Additives.

BIOLOGICAL DATA

Biochemical aspects

No information available.

Toxicological studies

Special studies on mutagenicity

Two different forms of polyethylenimine were evaluated for mutagenic activity in the presence or absence of an Archlor-induced rat liver activation system. The indicator organisms used were <u>Salmonella</u>

typhimurium strains TA1535, TA1537, TA1538, TA98, and TA100. A sample of polyethylenimine obtained prior to cross-linking with ethylene dichloride was mutagenic to strains TA1535 and TA100 both in the presence and absence of the activation system. The sample contained measurable amounts of ethylenimine which likely caused the mutagenic activity (Mortelmans & Shepherd, 1980).

An ethylene dichloride cross-linked sample, the form utilized in enzyme immobilization (Corcat P-600), did not contain measurable amounts of ethylenimine and was not mutagenic in any of the tester strains in the presence or absence of the activation system. At concentrations of up to 5000 micrograms/plate, 2 other forms of poly-ethylenimine, Corcat P-12 and Corcat P-18, were not mutagenic under the same test conditions (Mortelmans & Miron, 1981).

Polyethylenimine (P-1000) having a molecular weight of 70,000 was tested for mutagenicity with or without metabolic activation (source of the activating system not specified) using Salmonella typhimurium strains TA1535, TA1537, TA1538, TA98, and TA100 and E. coli strain wp2 uvrA. No mutagenic activity was found when concentrations of up to 5000 micrograms/plate were tested (Kajiwara et al., 1984).

Acute toxicity

Species	Route	LD_{50} (mg/kg b.w.)	Reference
Mouse	oral	2,800	BASF, 1959
Mouse	oral	8,000	Kobe University, 1974
Mouse	i.p.	40	BASF, 1959
Rat	oral (gavage in corn oil)		Rushbrook & Jorgenson, 1981
Male		7550[a]	
Female		7500[a]	
Both sexes	oral	2286[b]	
Male		2286[c]	
Female		1991[c]	

(continued)

Species	Route	LD$_{50}$ (mg/kg b.w.)	Reference
Rat	gavage (aqueous solution)	2000	Norris, 1973
Rat	oral	3,000	BASF, 1959
Rat	i.p.	70 mg/kg	BASF, 1959
Rabbit	oral	2,000 (no injury)	BASF, 1959
Rabbit	i.v.	4 (lethal dose)	BASF, 1959
Cat	i.v.	10 (lethal dose)	BASF, 1959

[a] Polyethylenimine type Corcat 600 (type used in enzyme immobilization)
[b] Polyethylenimine type Corcat P-12
[c] Polyethylenimine type Corcat P-18

It was not possible to determine the lethal oral dose of polyethylenimine in dogs and cats because of vomiting. Cats vomited within 30 minutes of receiving 100-500 mg/kg, so only a small amount of the material remained in the stomach. Doses of 100 mg/kg caused vomiting in dogs. In cases where vomiting did not occur, no adverse effects were reported (BASF, 1959).

In guinea-pigs, a single oral dose of 600 mg/kg was lethal to 6/10 animals and a dose of 800 mg/kg was lethal to 2/10 animals (BASF, 1959).

Short-term studies
Rats
Groups of 40 male and 40 female white rats received 0, 0.25, 0.5, or 1.0 g/kg b.w. of polyethylenimine in the diet for 8.5 months. No compound-related effects were reported with respect to mortality,

body weight, feed consumption, clinical signs, haematology, urinalysis, or absolute and relative organ weights. An elevation in blood alkaline phosphatase levels was observed in high-dose male rats as compared to the controls, but the values were within normal limits for the performing laboratory. No treatment-related changes were observed in gross and microscopic pathology studies carried out on 5 animals/sex/dose (McCollister & Copeland, 1968a).

Dogs

Groups of 4 male and 4 female beagle dogs were fed 0, 0.25, 0.5, or 1.0 g/kg b.w. polyethylenimine in the diet for 9 months. No compound-related differences were noted with respect to haematology, urinalysis, clinical chemistry, or bromsulfophthalein dye retention. Group mean body-weight gains were reduced as compared to controls in high-dose males and females. Relative liver and kidney weights were increased in high-dose females and relative kidney weights were slightly increased in high- and mid-dose males and mid-dose females.

The pathology report indicated that severe degenerative changes occurred in the kidney proximal convoluted tubules in the high-dose animals. The same lesion, but not as severe, was reported to occur in all mid-dose and 5 of 8 low-dose animals. No lesions of the proximal convoluted tubules were reported in control animals. All of the low-dose females and one of the low-dose males were reported to have the lesion to a "very slight" degree. Brown pigmentation of Kupffer's cells in the liver was considered to be compound-related. In the high-dose animals, it was present in 3 of 4 males at a moderate to very slight degree and in 2 of 4 females at a marked to moderate extent. In the mid-dose groups, 1 of 4 animals of each sex had the lesion present at a grade of "very slight". The lesion was not present in controls of either sex or low-dose males, but was present at a "very slight" grade in 1 of 4 low-dose females. There did not appear to be any other compound-related microscopic lesions (McCollister & Copeland, 1968b).

Rabbits

Rabbits were reported to tolerate several once-per-week oral doses of 1.0 g/kg polyethylenimine for several weeks without impairment of liver function. No injury to the liver or kidney was noted in rabbits receiving 1.0 g/kg daily, but the report stated that the rabbits tolerated a maximum of 6 doses. Daily doses of 0.5 g/kg were reportedly tolerated without any injury; however, details of the study were not provided (BASF, 1959).

Comments

Although not up to modern standards of toxicity testing, the 8.5-month rat study of free polyethylenimine did not reveal any adverse effects. In the 9-month dog study, compound-related changes in the kidney and liver were found. Since absorption and distribution studies are not available to show if appreciable uptake of this high-molecular-weight compound occurs, it is not clear what the mechanism of action is for the liver and kidney lesions. Polyethylenimine free of measurable levels of ethylenimine showed no mutagenic activity when tested with or without metabolic activation using Salmonella typhimurium strains.

EVALUATION

Estimate of acceptable daily intake for man

Polyethylenimine is considered to be a suitable substance for use as an immobilizing agent in the production of immobilized enzymes (see ethylenimine).

REFERENCES

BASF (1959). Preliminary report on the toxicity of PEI. Unpublished report of Badische Analin & Soda Fabrik AG. Submitted to WHO by UOP, Inc.

Kajiwara, Y., Oguru, S., & Takeyasu, K. (1984). Ames metabolic activation test to assess the potential mutagenic effect of polyethylenimine. Unpublished report of the Hita Research Laboratories, Chemicals Inspection & Testing Institute. Submitted to WHO by UOP, Inc.

Kobe University (1974). Acute toxicity of P-1000. Unpublished report of Kobe University, Medical Faculty, Public Health Section. Submitted to WHO by UOP, Inc.

McCollister, D.D. & Copeland, J.R. (1968a).   Results of 8.5 month
     dietary feeding studies of polyethylenimine in rats. Unpub-
     lished report of the Dow Chemical Company. Submitted to WHO
     by Cordova Chemical Company.

McCollister, D.D. & Copeland, J.R. (1968b). Results of 9 month dietary
     feeding studies of polyethylenimine in Beagle hounds.
     Unpublished report of the Dow Chemical Company. Submitted
     to WHO by Cordova Chemical Company.

Mortelmans, K.E. & Miron, K.L. (1981).   In vitro microbiological muta-
     genicity assays of Cordova Chemical Company's compounds
     Corcat P-12, Corcat P-18, and Corcat P-600.   Unpublished
     report of SRI International. Submitted to WHO by Cordova
     Chemical Company.

Mortelmans, K.E. & Shepherd, G.F. (1980).   In vitro microbiological
     mutagenicity assays of Cordova Chemical Company's compound
     PEI prepolymer, Sample No. SWM D32-084-1.   Unpublished
     report of SRI International. Submitted to WHO by Cordova
     Chemical Company.

Norris, J.M. (1973). Acute toxicological properties of PEI-600. Unpub-
     lished report of the Dow Chemical Company. Submitted to WHO
     by UOP, Inc.

Rushbrook, C.J. & Jorgenson, T.A. (1981).   Acute toxicity studies of
     three Corcat compounds.   Unpublished report of SRI
     International. Submitted to WHO by Cordova Chemical Company.

ETHYLENIMINE

EXPLANATION

Ethylenimine has not been previously evaluated by the Joint
FAO/WHO Expert Committee on Food Additives. Trace amounts of ethylenimine may potentially migrate into food from the presence of small
amounts of unreacted monomer present as a contaminant in polyethylenimine.

BIOLOGICAL DATA

Biochemical aspects

Distribution

Five male rats (Dow-Wistar strain) were injected i.p. with 0.3 to 0.4 mg/kg b.w. of ^{14}C-labelled ethylenimine and sacrificed after 24 or 96 hours. In both cases about 50% of the radioactivity was excreted in the urine and small amounts were present in faeces and exhaled air. A small amount of ethylenimine and a number of non-volatile metabolites were present in the urine and both ethylenimine and CO_2 were present in expired air. About 2.5% of the radioactivity was present in the liver after 24 hours and about 1% after 96 hours. Smaller amounts were present in many other tissues. The authors concluded that the compound was generally distributed throughout the rat before reaction with tissue components occurred (Wright & Rowe, 1967).

Excretion

Monoethanolamine and ethylenimine were excreted in the urine of rats following introduction of ethylenimine into the stomach (presumably by gavage). The urinary excretion of these 2 compounds accounted for approximately 50% of the administered dose. The fraction of the administered dose that was excreted in the urine was dose-dependent. At a dose of 1/20 the LD_{50} (0.85 mg/kg according to the authors) about 60% of the administered dose was excreted as monoethanolamine and ethylenimine in the urine over a 6-day period; 50% of the dose was excreted in the first 24 hours (Sanotsky et al., 1977).

Toxicological studies

Special studies on renal toxicity

Rats

Groups of 6 or 7 female Sprague-Dawley rats were given single subcutaneous injections of 0.25, 0.5, 1.0, 1.25, 2.0, 4.0, 6.25, or 8.0 mg/kg of ethylenimine in water. Animals were sacrificed 4 days after treatment. Renal papillary necrosis was observed at doses of 1.25 mg/kg ethylenimine and greater, whereas none occurred at 1.0 mg/kg and below (Axelson, 1978).

Rabbits

Renal medullary necrosis occurred in male and female New Zealand strain rabbits injected i.v. with a single 0.005 ml/kg dose (about 5 mg/kg) ethylenimine as a 1% v/v solution in water (Davies 1969; Davies, 1970).

Dogs

Groups of 4 male beagle dogs were given either a single i.v. injection of 3 μl/kg (about 3 mg/kg) ethylenimine or an initial injection of 0.6 μl/kg ethylenimine followed by a second injection of 1 μl/kg 3 days later. In the animals receiving 2 injections of ethylenimine the tubules of the collecting ducts and loop of Henle were dilated and inflamation was found in the renal pelvis. A small number of dilated tubules were observed in the medulla, and dilation of the distal convoluted tubules and collecting ducts of the cortex were noted. Almost complete necrosis of the papilla was observed in 2 dogs that became moribund after receiving the single high dose (3 μl/kg) of ethylenimine. Functional and clinical chemistry studies showed that treatment was related to impairment of renal function, proteinuria, and elevated excretion of urinary enzymes (Ellis et al., 1973).

Special studies on carcinogenicity

Mice

A carcinogenesis study of ethylenimine was carried out using 2 strains of mice, (C57BL/6 X C3H/Anf)F_1 (strain x) and (C57BL/6 X C3H/AKR)F_1 (strain y). Test animals were given the maximum tolerated dose (MTD), 4.64 mg/kg, daily by gavage in a 0.5% gelatin vehicle from day 7 through day 28 of age (the MTD for this study was the maximal dose giving no mortality when administered daily for 19 consecutive days). Thereafter, the animals received the same calculated daily dose of ethylenimine mixed in the feed. Groups of 18 animals/sex/strain were used. A number of other compounds were tested in this study. All the animals receiving a particular compound were placed in 1 of 4 rooms. Each room also contained an untreated control group of 18 animals/sex/strain. There was also 1 additional control group of 18 animals/sex/strain given a gelatin suspension during the time when the compounds were administered by stomach tube. Ethylenimine was actually employed as a positive control. There was at least 1 positive control group in each room. Ethylenimine-treated animals were on test for 77 or 78 weeks. The controls were on test for 78 to 89 weeks.

Data on mortality, body weight, food and water consumption, haematology, clinical chemistry, and non-carcinogenic pathological effects were not reported.

A list of the major organs examined microscopically was not provided. The report stated that all major organs and grossly-visible lesions were examined microscopically. However, the cranium was not dissected. Thyroid glands were sectioned in only 1 of the 5 control groups and not in the ethylenimine-treated animals. The tabulation of tumours listed only the following categories: hepatomas, pulmonary tumours, lymphomas, and total mice with tumours. The investigators lumped all 5 negative control groups together for purposes of comparison with positive controls and experimental compounds.

In the ethylenimine-treated animals there was a significant increase in the relative risk for development of hepatomas, pulmonary tumours, and total tumours. In strain x males given ethylenimine, the incidence of hepatomas, pulmonary tumours, and lymphomas was 15/17, 15/17, and 0/17, respectively. Corresponding values for the control strain x males were 8/79, 5/79, and 5/79. For strain x females given ethylenimine, the incidence of the different tumours was 11/15, 15/15, and 0/15, respectively, while the corresponding values for controls were 3/92, 3/92, and 5/92. In strain y males given ethylenimine, the incidence of hepatomas, pulmonary tumours, and lymphomas was 9/16, 12/16, and 0/16, respectively. Corresponding values for strain y male controls were 5/90, 10/90, and 1/90. In strain y females given ethylenimine, the respective tumour incidences were 2/11, 10/11, and 2/11; the corresponding values for strain y female controls were 1/82, 3/82, and 4/82. The actual dose of ethylenimine the animals received from the diet in this study is unclear. The concentration of ethylenimine in the feed was not measured and some material would certainly have been lost by volatilization and reaction with dietary constituents. An experiment, aimed at mimicking conditions of mixing in the diet and storage of ethylenimine during the conduct of the study, indicated loss of a considerable amount of the test substance

(Bionetics Research Labs, Inc., 1968; Green & Lowry, 1983; Innest et al., 1969).

The same group of workers gave single s.c. injections of 4.64 mg/kg b.w. of ethylenimine to groups of 18 male and 18 female mice of strains (C57B1/6 X C3H/Anf)F_1 and (C57BL/6 X C3H/AKR)F_1. The animals were then observed for 18 months. Tumours developed in 7 of the males of the (C57B1/6 X C3H/Anf)F_1 strain – 2 lymphomas, 2 hepatomas, and 5 pulmonary tumours. In the (C57BL/6 X C3H/AKR)F_1 strain, 6 of 18 males each developed a lung tumour. For males of both strains the total number of tumours and the incidence of pulmonary adenomas was significantly greater than in the controls (P < .01). In the treated females, 1 animal of each strain developed a lung tumour. The controls consisted of 9 separate groups lumped together for purposes of comparison with a total of about 160 animals/sex/strain (Bionetics Research Labs., Inc., 1968).

Fragmentary data are available from another study of ethyl-enimine administered s.c. at doses of 0.4, 1.3, or 4.0 mg/kg at weekly intervals for 48 weeks to 187 male and female C56BL X CBA mice. After 2 years, dosed animals had an increased incidence of sarcomas at the injection site, tumours of the harderian gland and lung, and malignant hepatomas (Linnik, 1980).

Rats
A series of ethylenimine derivatives was administered by s.c. injection to groups of 6 male and 6 female albino rats. Ethylenimine was injected twice weekly in an arachis oil vehicle. The dosing regimen occurred over a 67-day period, with a total dose of 20 mg/kg being administered. Sarcomas were found at the injection site in 5 of 6 males and 1 of the females. The tumours were discovered between 355 and 511 days after the beginning of dosing. The study was termi-nated at 546 days after the beginning of dosing. No tumours were found at locations remote from the point of injection. In controls injected with arachis oil for the same duration as the ethylenimine group, 1 of 10 males developed an injection-site sarcoma at 568 days while 2 tumours developed at remote sites. None of 9 control females developed

sarcomas at the injection site; however, a fibroma was found in 1 animal at that site and tumours were found in 2 other animals at remote sites.

In another experiment, groups of 6 males and 6 females were given ethylenimine dissolved in water twice weekly by s.c. injection. A total dose of 10 or 12 mg/kg was given to males and females, respectively, over a 59-day period (dosed 5 days/week). The animals were observed up to 540 days; 2 sarcomas were observed at the injection site in females and none in males, while a transitional cell carcinoma of the kidney was seen in 1 male. Concurrent control groups injected with a carbowax-300 vehicle had no injection-site sarcomas (Walpole et al., 1954).

Special studies on mutagenicity

Sex-linked recessive lethals and translocations were reported in a study in which day-old male Drosophila were treated with an i.p. injection of 0.4 μl of a 10^{-2} M solution of ethylenimine and subsequently mated. The authors stated that the compound was radiomimetic in producing chromosomal abnormalities – translocations were nearly as frequent as sex-linked lethals – but resembled some chemical mutagens, such as mustards, in producing delayed lethals and translocations which appeared in the F_2 generation from F_1 parents that appeared normal (Alexander & Glanges, 1965).

Other studies have also found that ethylenimine induces recessive lethals and translocations in Drosophila (Lim & Snyder, 1968; Alexander, 1967).

Ethylenimine was mutagenic in Salmonella typhimurium strains TA1535 and TA100 without metabolic activation (McCann et al., 1975).

The compound was also reported to be mutagenic in spot tests on Neurospora crassa strain N-23 (reverted by base pair mutagens), but not by strain N-24 (reverted by frame shift mutagens) (Ong, 1978).

Severe inhibition of replicon initiation and blocked-chain elongation occurred when 5×10^{-4} M ethylenimine was added to cultured HeLa 53 cells (Painter, 1978).

Injection of 1 mg/kg ethylenimine (i.p.) into male mice resulted in about a 2/3 reduction in incorporation of ^{3}H-thymidine into testicular DNA as compared to control animals (Seiler, 1977).

Lymphocyte cultures from 10 workers exposed to workroom ambient-air concentrations of 0.5 ppm ethylenimine reportedly did not show an increase in chromosomal aberrations (Gaeth & Thiess, 1972).

The effect of incubation with ethylenimine was studied in cultures of WI-38 cells and in leukocytes from an adult male volunteer. A concentration of 10^{-2} M ethylenimine was cytotoxic to the WI-38 cells. Concentrations of 10^{-3} and 10^{-4} M were associated with chromatid breaks, gaps, and exchanges. Radiolabel studies with leukocyte cultures incubated with 10^{-4} M ethylenimine also showed gaps, exchanges, and breaks, occurring primarily in the S-period of the cell cycle (Chang & Elequin, 1967).

Acute toxicity

Species	Route	LD$_{50}$ (mg/kg b.w.)	Reference
Rat	oral	17	Santoski et al., 1977
Rat	oral	15	NIOSH, 1977
Rat	i.p.	3.8	NIOSH, 1977

Short-term studies
No information available.

Long-term studies
No information available.

Comments

Ethylenimine appeared to be carcinogenic when administered orally to 2 strains of mice. Both sexes were affected, with the liver and lung being the major target organs. When administered s.c. to rats and mice, the compound appared to be associated with sarcomas at the injection site as well as at other locations. Ethylenimine was found to be mutagenic in Neurospora, Salmonella, and Drosophila, and chromosomal aberrations occurred in cultured mammalian cells exposed to the compound.

Use of the oral feeding study in mice for risk analysis is complicated because of the one-dose design of the experiment, questions about the actual dose of ethylenimine received by the animals, the appearance of tumours at 2 sites, and the very high tumour incidence. While clearly indicating that ethylenimine is carcinogenic, the mouse study is not up to modern standards of toxicity testing. Carcinogenicity studies by the oral route are available only in the mouse. Reproduction and teratology studies are not available.

EVALUATION

Level causing no toxicological effect

Ethylenimine has been determined to be carcinogenic in mice. A "no-effect" level in experimental animals has not been established.

Estimate of acceptable daily intake for man

Acceptable on condition that human exposure to ethylenimine as a result of its migrations into food from immobilized enzyme preparations is reduced to the lowest level technically possible (see polyethylenimine).

REFERENCES

Alexander, M.L. (1968). Mosaic mutations induced in Drosophila by ethylenimine. Genetics, 56, 273–281.

Alexander, M.L. & Glanges, E. (1965). Genetic damage induced by ethylenimine. Proc. Nat'l Acad. Sci., 53, 282–288.

Axelson, R.A. (1978). Experimental renal papillary necrosis in the rat: the selective vulnerability of medullary structures to injury. Virchows Arch. A. Path. Anat. and Histol., 381, 79-84.

Bionetics Research Labs., Inc. (1968). Evaluation of carcinogenic, teratogenic, and mutagenic activities of selected pesticides and industrial chemicals. Volume I, Carcinogenic Study. Prepared for National Cancer Institute. Available from National Technical Information Service, U.S. Department of Commerce.

Chang, T. & Elequin, F.T. (1967). Induction of chromosome aberrations in cultured human cells by ethylenimine and its relation to cell cycle. Mut. Res., 4, 83-89.

Davies, D.J. (1969). The structural changes in the kidney and urinary tract caused by ethylenimine (vinylamine). J. Path., 97, 695-703.

Davies, D.J. (1970). The early changes produced in the rabbit renal medulla by ethylenimine: electron-microscope and circulatory studies. J. Path., 101, 329-332.

Ellis, B.G., Price, R.G., & Topham, J.C. (1973). The effect of papillary damage by ethylenimine on kidney function and some urinary enzymes in the dog. Chem-Biol. Interact., 7, 131-141.

Gaeth, V.J. & Thiess, A.M. (1972). Chromosome studies on chemical workers, Zentralbl. Arbeitsmed. Arbeitsschutz, 22, 357-362.

Green, D.R. & Lowry, J.R. (1983). Stability of ethylenimine in mouse chow and 0.5% gel solution. Unpublished report of Cordova Chemical Company. Submitted to the World Health Organization by Cordova Chemical Company.

Innes, J.R.M., Ulland, B.M., Valerio, M.G., Petrucelli, L., Fishbein, L., Hart, E.R., Pallota, A.J., Bates, R.R., Falk, H.L., Gart, J.J., Klein, M., Mitchell, I., & Peters, J. (1969). Bioassay of pesticides and industrial chemicals for tumorigenicity in mice: A preliminary note. J. Nat. Cancer Inst., 42, 1101-1114.

Lim, J.K. & Snyder, L.A. (1968). The mutagenic effects of two mono-functional alkylating chemicals on mature spermatozoa of Drosophila. Mutation Res., 6, 129–137.

Linnik, A.B. (1980). Study of the carcinogenic effect of ethylenimine on F_1(C57BL X CBA) mice. Eksp. Onkol., 2, 67–68.

McCann, J., Choi, E., Yamasaki, E., & Ames, B.N. (1975). Detection of carcinogens as mutagens in the Salmonella/microsome test: Assay of 300 chemicals. Proc. Nat'l Acad. Sci., 72, 5135–5139.

NIOSH (1977). Registry of toxic effects of chemical substances. National Institute of Occupational Safety and Health, Washington, DC, USA.

Ong, T. (1978). Use of the spot, plate and suspension test systems for the detection of the mutagenicity of environmental agents and chemical carcinogens in Neurospora crassa. Mutation Res., 53, 297–308.

Painter, R.B. (1978). Inhibition of DNA replicon initiation by 4-nitroquinoline 1-oxide, adriamycin and ethylenimine. Cancer Res., 38, 4445–4449.

Sanotosky, I.V., Muravieva, S.I., Zaeva, G.N., & Semiletkina, N.N. (1977). Urinary excretion of ethylenimine and its metabolite, monoethanolamine, under experimental conditions. Gig. Tr. Prof. Zabol., p. 10–14.

Seiler, J.P. (1977). Inhibition of testicular DNA synthesis by chemical mutagens and carcinogens. Preliminary results in the validation of a novel short-term test. Mutation Res., 46, 305–310.

Walpole, A.L., Roberts, D.C., Rose, F.L., Hendry, J.A., & Homer, R.F. (1954). Brit. J. Pharmacol., 9, 306–323.

Wright, G.J. & Rowe, V.K. (1967). Ethylenimine: Studies of the distribution and metabolism in the rat using carbon-14. Toxicol. Applied Pharmacol., 11, 575–584.

FLOUR TREATMENT AGENT

CHLORINE

EXPLANATION

Chlorine is used for the treatment of flour for special purposes, such as cake manufacture. Chlorine, as a flour treatment agent for special-purpose flour, was reviewed at the ninth meeting of the Committee (Annex 1, reference 11) but an acceptable level of use was not established. At the earlier review, it was concluded that "long-term studies using appropriate products made from flour treated with chlorine at various levels will be needed".

Since the previous review, further studies have become available and are summarized and discussed in the following monograph.

BIOLOGICAL DATA

Biochemical aspects

Interaction of chlorine with flour

Treatment of flour with 1950 ppm of chlorine resulted in a lowering of the unsaturated fatty acids in the flour to 40% of levels in untreated flour. Oleic acid was probably converted into dichloro-stearic acid while a range of chlorinated compounds was formed from linoleic and linolenic acids. Treatment of flour with up to 120 ppm of chlorine did not materially change the major fatty acids (Coppock et al., 1960; Daniels, 1960).

When soft-wheat flours were treated with chlorine, the chlorine content of the lipids was markedly increased, that of water-soluble components to a lesser extent, and of the gluten only slightly. The lipids and water-soluble components comprised only 5% of the flour, but contained more than 90% of the added chlorine. The chlorine-containing lipids showed a decreased iodine value (Gilles et al., 1964).

The chlorine content of untreated flour was found to be 430-540 mg chlorine/kg flour and of bleached flour 1310-1890 mg chlorine/kg flour. Almost all of the additional chlorine was in water-solubles and gluten (including lipid), and treatment did not significantly increase the chlorine in the prime starch. At least 50% of the additional chlorine in the gluten was in the lipid fraction (Sollars, 1961).

Absorption, distribution, and excretion

Rats fed diets containing 4.1% lipids extracted from chlorine-treated flour showed a decrease in polyunsaturated fatty acids and a corresponding increase in palmitic, oleic, and palmitoleic acids in fat depots when compared with controls receiving 4.1% lipid from untreated flour. The chlorine content of adipose tissue was, however, only slightly increased. Effects of this nature could not be demonstrated at lower levels of dietary lipid more comparable with the amounts that might be ingested from treated flour (Daniels et al., 1963).

Oleic, linoleic, and linolenic acids, and their triglycerides, were chlorinated with ^{36}Cl-chlorine and administered orally to male Wistar rats. Tritiated parent fatty acids acted as controls for comparisons of absorption, tissue distribution, and excretion of the compounds. Absorption of the chlorinated compounds was greatly reduced compared to the parent compounds and tissue deposition was generally greatly reduced, although deposition of chlorinated oleic acid in the heart was similar to that of oleic acid and deposition of chlorinated triolein was greater than that of unchlorinated triolein. Water-soluble chlorination products were readily absorbed, were not incorporated into body lipids or proteins, and were rapidly excreted in the urine (Cunningham & Lawrence, 1977a; 1977b).

Single oral doses of tritiated oleic acid of ^{36}Cl-chlorinated oleic acid were administered to groups of male rats which were then killed at periods of 1-28 days after administration. The chlorinated oleic acid was absorbed to a lesser extent than ^{3}H-oleic acid (72.3% compared with 91.5%). The blood-brain barrier appeared effective against transfer of chlorinated oleic acid, but the compound was widely distributed throughout other tissues. The liver appeared to be the main site for dechlorination of the chlorinated oleic acid, the half-lives in the organs being 19.5 days in the kidney, 10.7 days in the liver, 10.0 days in the brain, 8.3 days in the heart, and 5.1 days in blood (Cunningham & Lawrence, 1976).

Placental and mammary transfer of ^{36}Cl-chlorinated oleic acid and tritiated oleic acid were studied after oral administration to rats. Transfer across the placenta of the chlorinated acid occurred at half the rate of the unchlorinated acid and 0.4% of the dose administered remained in the foetus after 19 days. Mammary transfer of chlorinated oleic acid to a 2-week-old suckling rat occurred to the extent of 15.9% of the parental dose in 24 hours, compared with 34.0% for unchlorinated oleic acid. The corresponding transfer of chlorinated linoleic and linolenic acids accounted for only 2.4% and 2.7% of the dose administered, respectively (Cunningham & Lawrence, 1977c).

Toxicological studies
Special studies on reproduction

Rats

A multigeneration study was performed in which male and female Fischer rats received diets containing cake made from flours treated with 0, 1000, 1500, or 2500 ppm chlorine, dried to a moisture content of 6%, and incorporated into the diet at a level of 75%. An additional control group received commercial Purina chow. The study was continued through 3 successive generations. Selected animals of the F_{1a} generation were used for long-term studies (see below) and the F_{3b} litter of the cake control and top-treatment groups were used for a teratology study (see below). No adverse effects were seen in the F_{1a} generation. In the F_{2a} generation there were decreased numbers of live births in the highest-treatment group and an increased mating-parturition time in the mid- and highest-treatment groups (gestation time was not determined). Weanling weights decreased with increasing levels of chlorination, though weanling weights of all groups receiving cake were greater than those of the Purina chow control group. In the F_{3a} generation, no effects attributable to treatment were observed; there was a non-treatment related decrease in the incidence of live births in cake-fed animals (Gumbmann & Gould, 1979c).

Female rats were fed a diet in which 93% of the total nutrients were provided by cake made from flour chlorinated at a level of 10,718 ppm. They were mated with males receiving similar diets or unchlorinated cake diets. Each female produced 2 litters. The results, though poorly presented, indicated no adverse effects on the reproduction parameters monitored, viz: Number of litters, live and dead pups, and those surviving to weaning. However, a slight decrease in pup weaning-weight was noted. As a comparison, mating was performed between control males and females, and between males and females receiving 2200 ppm chlorinated cake diets. No differences were noted in reproduction parameters, with the exception of a slight reduction in pup weaning-weight (Fisher et al., 1979).

Five groups, each containing 3 rats, were fed diets containing 0, 0.82, or 4.1% untreated flour lipid, or 0.82 or 4.1% lipid from flour treated with chlorine at a level of 1950 ppm. At the 4.0% level, the groups receiving lipid from the treated flour had thinner and rougher fur, fertility was reduced, and lactation was less efficient than in rats from the control group. Similar depression of fertility and lactation was observed with the group receiving 0.82% treated lipid. These effects were observed consistently through 4 generations and were not relieved by feeding 2% linoleic acid for 8 weeks (Daniels et al., 1963).

Special study on teratogenicity
Forty-eight male and 48 female Fischer rats from the cake control and highest-treatment groups of the F_3 generation in the multigeneration study (see above) were mated and the foetuses removed

by Caesarian section on day 20 of gestation. Numbers of corpora lutea, implantation and resorption sites, and live foetuses were recorded. After gross examination, sexing and weighing, half the foetuses were examined for skeletal defects and half for soft tissue abnormalities. Pregnancy rates in both groups were low, approximately 50%, and litter sizes were decreased compared with F_0 and F_1 matings. A slight reduction in ossification was noted in the chlorine-treated cake group, but no treatment-related soft tissue malformations were observed (Gumbmann & Gould, 1979c).

Acute toxicity
No information available.

Short-term studies
Rats

Two groups of 11 rats were fed for 16 days on dried cake made from a commercial cake flour treated with chlorine at a level of 0 or 1563 ppm. The cake constituted 90% of the final diet, to which was added vitamin mix in sugar and a 4% mineral mixture. No significant differences in food intake or growth rate were observed (FMBRA, 1968).

Four groups of 10 male weanling rats were fed for 29 days on dried or undried cake flour treated with chlorine at a level of 0 or 1563 ppm; the dried flours were supplemented with 0.15% lysine. No significant differences in growth rates were observed (FMBRA, 1968).

In preliminary trials, small groups of weanling rats were fed for 16-95 days on diets in which 93% of the total nutrients were provided by cake prepared by several different recipes using flour chlorinated at commercial levels of 1100 or 2200 ppm or at higher levels of chlorination, 5359 or 10,718 ppm. Food consumption and growth rates were recorded; post-mortem findings were limited to organ weights, analyses of perirenal fat, haemoglobin concentrations, and PCV.

At chlorination levels up to 5359 ppm, growth rates were unaffected, but at 10,718 ppm growth rates and food consumption were reduced, probably due to reduced palatability. As an additional study, the effects on growth rate were monitored using cake from chlorinated

flour giving the required chlorination level either by direct chlorination or by dilution of a more highly-chlorinated flour with untreated
flour. The results showed that the growth rate was decreased proportionately after feeding increasing levels of cake made with "diluted"
flour (and with a similar corresponding reduction in food consumption)
whereas "direct" chlorination to an equivalent level resulted in little
or no reduction in growth rate at chlorination levels up to 5000 ppm.
Post-mortem findings revealed no treatment-related effects at a level
of chlorination of 2200 ppm, but at 5359 and 10,718 ppm absolute liver
and kidney weights were significantly increased and perirenal fat
decreased, especially in females. PCV and haemoglobin concentrations
tended to be lower in treated animals, sometimes significantly, but the
levels were always within the expected range (Fisher et al., 1979).

Flours chlorinated to levels of 2000 ppm or 10,000 ppm were
fed to male Wistar rats at a level of 87.4% of the diet. Severe growth
retardation was reported after 2 weeks, accompanied by increased liver
weights. Lipids extracted from these flours had similar effects as did
rat chow diets containing 2000 or 6000 ppm chlorine in the form of
chlorinated gluten. A rat chow diet containing 2000 ppm chlorine as
chlorinated flour lipids increased absolute weights of the liver by
40%, kidney by 20%, and heart by 10% compared with pair-fed controls.
Histological examination of the livers showed hepatocytes with coarse,
foamy reticulated cytoplasm which were "collapsing and rupturing".
Animals fed chlorinated gluten showed hepatocytes "with a glassy
appearance" (Cunningham et al., 1977).

The ether-extracted lipids of flour were chlorinated using
chlorine levels required to chlorinate the parent flour to 2000 ppm or
10,000 ppm. The lipid and the extracted-flour residue were
incorporated into otherwise identical diets fed to male rats for 2
weeks in the following combinations: chlorinated lipid + unchlorinated
residue; chlorinated residue + unchlorinated lipid; or, unchlorinated
residue + unchlorinated lipid (control). Body-weight gains were
reduced by the chlorinated lipid, but not by the chlorinated residue.
Relative organ weights were affected as follows: liver and heart
weights were increased both by chlorinated lipid and by chlorinated

residue; brain weights were increased by chlorinated residue, but not by chlorinated lipid. Liver lipids (% w/w) were increased by chlorinated residue, but not by chlorinated lipid (Cunningham & Lawrence, 1978).

In a 2-week feeding study, male rats were pair-fed on diets containing the ether-extracted lipids obtained from a cake made from chlorinated flour (2000 ppm) or from unchlorinated flour. The lipid was incorporated into ground rat chow at a level of approximately 6%. The body weights of animals fed the treated diet were lower than those fed the control diet by about 5% ($p < 0.05$), liver weights were higher by 12% ($p < 0.01$), and brain weights were lower by 4.5% ($p < 0.01$). Kidney and heart weights and liver lipids (% w/w) did not differ significantly (Cunningham & Lawrence, 1978).

Ten-week feeding studies in rats were carried out using flour chlorinated to 0, 2000, or 10,000 ppm at dietary levels of 87.4%. Ether-extracted wheat gluten was directly chlorinated to 0, 20,000, or 50,000 ppm and incorporated at a level of 10% in a rat-chow diet, which was also used in these 10-week studies.

In the chlorinated-flour experiment, weight gains were significantly reduced by the treatment ($p < 0.05$ and $p < 0.01$ for the 2000 and 10,000 ppm chlorination levels, respectively). Relative weights of the kidneys, liver, heart, and brain were significantly increased ($p < 0.01$) compared to controls at the 10,000 ppm treatment level; relative kidney weights were also increased at the 2000 ppm treatment level. At the latter level, heart and brain relative weights were similar to controls.

Male rats fed diets containing gluten chlorinated to 50,000 ppm (5000 ppm in the diet) showed significantly-reduced weight gains compared with controls, but this was not the case when the dietary level was 2000 ppm. Relative kidney weights were increased ($p < 0.01$) at both treatment levels, and relative brain weights were increased ($p < 0.01$) only at the 5000 ppm dietary level (Cunningham & Lawrence, 1978).

Flour was chlorinated to levels of 1250 or 2500 ppm and fed to groups of Wistar rats at 80% of the diet for 28 days. Body weights were recorded daily and, at termination, the liver, kidneys, heart and brain from each animal were weighed and subjected to histological exam- ination. Rates of growth and food intake were only slightly reduced by treatment, particularly in females. Absolute organ weights did not differ significantly from controls, whereas treatment-related increases in relative liver and kidney weights were observed in both sexes. Histological changes observed were confined to vacuolation in the liver (caused by glycogen accumulation) and renal calculi in the collecting tubules, especially in females, but these effects were independent of the level of chlorination (Fisher et al., 1979).

Dogs

Two groups of 4 or 6 dogs were fed diets containing flour treated with 356 ppm chlorine for 21-38 days. This diet did not cause "running fits". Dogs which had developed "running fits" on agene-treated flour recovered on being switched to chlorine-treated material (Arnold, 1949; Radomski et al., 1948; Bentley et al., 1948; Newell et al., 1947).

Groups of 3 male and female beagle dogs were fed cake made from flour treated with 0, 1000, 1500, or 2500 ppm chlorine, dried to a moisture content of 6%, and incorporated into diets at a level of 75% for 6 months. Appearance, behaviour, and food intake were monitored daily; body weights were recorded weekly. Blood samples taken prior to treatment and at termination were subjected to haematological examin- ation (RBC, total and differential white cell and platelet counts, haemoglobin, PCV, and prothrombin time) and serum analysis (SAP, SGOT, SGPT, OCT, albumin, protein, cholesterol, bilirubin, BUN, thyroxine, Na, K, and Cl). Urine samples taken prior to treatment and at termin- ation were examined for colour, pH, specific gravity, and sediment. All animals were autopsied, organ weights were recorded, and tissues were subjected to histopathological examination.

No treatment-related effects on appearance, behaviour, growth, or food intake were observed. One mid-dose female developed peritonitis after 5.5 months and was killed and autopsied. No adverse effects on haematological parameters were observed. Serum analyses revealed a treatment-related decrease in thyroxine levels in females, which was statistically significant in the high-dose group only; other parameters examined showed no treatment-related differences. Urine composition was unaffected by treatment. No treatment-related effects were noted in the weights of the brain, testes, spleen, adrenals, thyroid, or pituitary; relative kidney weights were decreased in high-dose males and there was a non-dose-related increase in absolute and relative left ventricle weight and absolute heart weight in females. The main pathological changes seen were mild renal medullary mineralization in all dogs and patchy granulomatous pneumonia, but the effects were not treatment-related (Gumbmann & Gould, 1979d).

Long-term studies
Mice

Groups of 60 male or 60 female 4-5 week-old mice, Theiller's original strain, were fed diets containing cake made from flour treated with 0 (cake controls), 1250, or 2500 ppm chlorine that had been dried to a moisture content of 12.6% and incorporated into the diets at a level of 79% w/w. Groups of 30 male or 30 female animals were fed commercial diet 41B (41B controls). Satellite groups (30 animals of each sex) were used for interim examinations at 3 and 12 months for haematology, serum analyses, and urinalyses (12 months only). The intended duration of the study was 80 weeks but, due to poor survival, the study was terminated at month 16 for males and month 17 for females.

Cage weights (5 mice/cage) and food consumption were recorded weekly for the first month and then monthly thereafter. Haematological examinations (haemoglobin, RBC, total and differential leucocyte counts, and platelet counts) were performed on 10 mice/sex/group during months 3, 12, and at termination. Serum analyses (BUN, glucose, SGOT, SGPT, SAP, and total protein) were performed on 5 mice/sex/group at months 1 and 12 and on 10 mice/sex/group at termination. Pooled urine samples from 10 mice/sex/group were subjected to urinalysis (volume, pH, SG, protein, glucose, ketones, bile pigments, urobilinogen, blood pigments, and renal concentration tests) at month 12 and at termination.

At termination, all animals were subjected to gross necropsy and histopathological examination, the latter being limited to abnormal tissues, including all masses, brain, heart, lung, liver, spleen, kidneys, uterus, and gonads. Renal adipose tissue samples were analysed for covalent chlorine.

No adverse treatment-related effects were seen on appearance, behaviour, food intake, or growth, but all the cake-fed animals had a reduced food intake relative to 41B-diet controls during the first 6 months of the study. Body-weight gain was markedly higher in animals receiving cake diets (chlorinated or not) compared with 41B-diet controls, and between months 6 and 12 the former groups weighed nearly twice as much as the latter. The energy value of the cake diet was 3.6 kcal/g and that of the 41B diet was 3.0 kcal/g.

The mortality rate among males fed the cake diets was similar in all groups, but in females the groups receiving chlorinated cake showed a higher mortality rate than the cake controls. However, the mortality rate of mice in all cake-fed groups was greater than the mortality rate of the 41B control mice after 12 months and much higher than expected for the strain of mice used.

Statistically-significant treatment-related reductions in red cell counts were seen in cake-fed males at 3 and 12 months; increases in white cell counts were observed at month 3; a non-treatment-related increase in haemoglobin was observed at month 12. No significant differences in haematological parameters were seen among cake-fed females. Compared with 41B controls, however, the following differences were seen: at month 12, decreased haemoglobin in all male and female groups; decreased red cell counts in high-treatment males and all females, and decreased PCV in all females. At termination, all male values were similar, but reduced Hb, PCV and RBC persisted in females. Despite the observed differences, none were below normal ranges and were not related to the level of chlorination. Clinical chemical analysis at months 1 and 12 revealed similar values between all cake-fed groups except for a decrease in BUN in females of the high-treatment group at month 1. Terminal analyses revealed a decreased BUN in males of the high-treatment group, decreased SGPT and total protein in females of this group, and a dose-related decrease in serum alkaline phosphatase in females. Comparison with 41B controls revealed increases in BUN at month 1 (all cake-fed males and cake controls and 1250 ppm females) and in all females at month 12 and at termination. No treatment-related effects were seen in urinalyses nor in organ weights, but comparison with 41B controls revealed significant increases in absolute and relative weights of the heart, liver, kidneys, and spleen in both sexes.

At autopsy, no macroscopic changes attributable to chlorination of the cake flour were observed, but all cake-fed mice showed an increased incidence of excess adipose tissue. The predominant histopathological change seen in all groups was amyloid deposition, principally in the spleen, liver, kidneys, heart, ovaries, and uterus, but the incidence was not treatment-related. In the kidneys, there was a higher incidence (not significant) of glomerulonephrosis in cake-fed mice compared to 41B controls (35–50% compared with 25%, respectively), a significant increase in the number of calcareous deposits in the medulla or low in the tubules of the pelvis in the high-treatment group (though the amounts deposited were minimal), and a high incidence of cystis glomeruli in all cake-fed mice. In the heart, a high incidence of calcareous deposits was noted in cake-fed mice only, but these deposits were not correlated with the degree of glomerulonephrosis. Tumour incidence was similar in all groups; reticulum cell neoplasms (maximum of 1/group) were observed in all male groups and in the female high-treatment group.

Analysis of renal adipose tissue revealed a treatment-related deposition of fat containing covalently-bound chlorine.

The changes observed in this study could mostly be attributed to the nature of the diets fed to the mice rather than to the level of chlorination (Fisher et al., 1979; Ginocchio et al., 1983).

Groups of 40 male and 40 female CDI mice were fed diets containing cake made from flour treated with 0, 1000, 1500, or 2500 ppm chlorine, dried to a moisture content of 6%, and incorporated into diets at a level of 75%. An additional group (Purina controls) was fed commercial Purina chow diet. The duration of treatment was 85 weeks for cake-fed males, 93 weeks for Purina controls and low-treatment females, and 103 weeks for the remaining females. Appearance and behaviour were monitored daily, while cage weights and food intake were

recorded weekly for 28 weeks and monthly thereafter. Haematology
analyses (RBC, total and differential leucocytes, PCV, and Hb) were
performed on 8 mice/group at 7, 12, 15, and 18 months and at termination. Serum analyses (albumin, total protein, SAP, and BUN) were
carried out after 15 months and at termination. Urinalysis (5 mice/
group) was performed at the same times as the blood samples were
analysed.

At 15 months, 5 mice/sex/group were sacrificed, and these
animals and those killed at termination were autopsied. Liver, spleen,
kidney, heart and testes weights were recorded and all animals were
subjected to histopathological examination.

No adverse treatment-related effects were recorded for
appearance, behaviour, or growth rate. Survival was unrelated to the
level of chlorination and was slightly greater in all cake-fed groups
than in the Purina controls. No adverse effects were seen on haematological parameters and no differences attributable to chlorination
were seen in serum analyses or urinalyses. At interim kill (15
months), increased absolute and relative kidney and heart weights were
recorded in males of the 2500 ppm group and increased absolute and
relative liver weights were seen in both sexes of the 1000 and 1500 ppm
groups; at termination, no differences were noted. The main pathological changes, none of which could be attributed to chlorination,
were: a high incidence of amyloidosis in all groups (less marked in
females), increased reticuloendothelial hyperplasia in males of the
2500 ppm group, and acute inflammation of the urinary tract in males of
the 1500 and 2500 ppm groups.

Although the overall tumour incidence was unaffected by
chlorination, there was a significantly-increased incidence of lympho-
haematopocitic tumours in mid- and high-treatment females compared with
the Purina controls, but not compared with the cake controls; the
tumours were classified as malignant lymphoma. Actuarial analyses of
the lymphomas revealed that the differences between the numbers of
tumours expected and observed were not significant. There did not
appear to be any relationship with the level of chlorination; however,
the apparent induction period for tumour formation was decreased
(Gumbmann & Gould, 1979a).

Rats

Groups of 60 male or 60 female Wistar-derived rats were fed diets containing cake made from flour treated with 0, 1250, or 2500 ppm chlorine dried to a moisture content of 12.6% and incorporated at a level of 79% into their diets for 104 weeks. An additional group received commercial diet 41B (41B controls). Satellite treatment groups were also included for interim studies. Appearance and behaviour were monitored daily, food consumption and body-weight gains were recorded weekly for the first 13 weeks and monthly thereafter, and animals were palpated for tumours when body weights were recorded. Water consumption was determined at monthly intervals. Blood samples were taken from 10 rats/sex/group for haematological examination (Hb, PCV, RBC, and total and differential leucocytes) at 3, 12, 18, and 24 months and from 5 rats/sex/group for serum analyses (BUN, glucose, SGOT, SGPT, SAP, total protein, and albumin) at 3, 12, and 18 months; serum analyses were also performed on 10 rats/sex/group plus all female survivors at 24 months. Urinalyses (pH, SG, protein, glucose, ketones, blood and bile pigments, urobilinogen, NAG, and creatinine) were carried out on samples from 10 rats/sex/group at 6, 12, 18, and 24 months; renal urine concentration tests were carried out in 5 rats/sex/ group at 6-month intervals, urinary cell counts at month 18, and urinary GOT at 18 and 24 months.

At termination, all animals were autopsied and weights of the brain, heart, adrenals, liver, spleen, gonads, pituitary, kidneys, thyroid, and uterus were recorded. Because few females survived the 2-year period, organ-weight analyses were performed on the 5 rats/group satellite-treatment animals killed at month 18 for serum analysis. All animals were subject to gross pathological examination and detailed histopathology.

At an early stage in the study, evidence of respiratory distress was noted in several rats, particularly the 41B controls. Serological evidence of sialodacryadenitis was shown in animals that died. Because respiratory distress appeared to be exacerbated by the powdered nature of the 41B diet, this diet was fed in pellet form from week 19 onward.

There were no differences in food intake among cake-fed groups, but those fed the 41B diet had a higher food intake (20-25%

higher), especially after the diet was changed to pellet form; the
energy value of the cake-based diets was 3.5 kcal/g and that of the 41B
diet was 3.0 kcal/g. Rate of weight gain was similar in all cake-fed
rats, irrespective of whether the cake flour was chlorinated, and 6-7%
higher than in the 41B controls. Water intake was unaffected by
treatment.

　　　　Mortality was similar in all cake-fed groups, including
cake-fed controls. Initially, in male 41B controls mortality was much
higher than in rats fed the cake-based diet (due to early loss from
respiratory illness), but was similar to other groups by week 80. In
females, mortality of 41B controls was lower than that of cake-fed
animals throughout the study. After 104 weeks, mortality ranged from
40% (41B controls) to 63% (cake controls) in males, and 85% (41B controls) to 98% (cake controls and 1250 ppm group) in females.

　　　　Isolated changes relative to cake controls were seen in some
haematological parameters, but there were no consistent treatment-
related effects. Compared with cake-fed controls, there were
treatment-related increases in SGOT and SGPT at month 12 in males, but
SGPT levels were within the normal range. At month 24, total serum
protein was decreased in high-treatment males.

　　　　No treatment-related effects on urine composition were
noted. Blood-stained urine was noted in the floors of a few cages from
each male group, but it did not appear to be treatment-related; only
one cage of females was affected. The ratio of NAG to creatinine
revealed no treatment-related effects, even in rats with kidney lesions
observed at post-mortem.

　　　　No treatment-related effects on organ weights were noted at
termination. However, taking all females (including 5 animals killed
at month 18) into account, there was a significant reduction in spleen
weights relative to cake-fed controls, though the spleen weights of all
cake-fed females were higher than those of 41B controls.

　　　　The main gross pathology finding was an increased incidence
of enlarged kidneys in rats receiving the cake-based diets. Histology
revealed glomerulonephrosis of varying severity in more than 90% of the
animals, to a greater extent in those fed cake-based diets. Renal
calculi were observed mostly in animals fed cake diets (80% of females,
15% of males) and few 41B rats were affected (6 females, no males).

There was an increased incidence of haematopoiesis in the spleens of females fed cake-based diets.

Pituitary chromophobe adenomas were more common in the female groups (approximately 65% of each group was affected) than in the male groups (15-33%), and probably contributed to early mortality. However, no differences were noted in tumour incidence that could be attributed to chlorination of cake flour.

Perirenal adipose tissue from 5 rats/sex/group showed covalently-bound chlorine levels related to the level of flour treatment (Fisher et al., 1979).

Groups of 40 male or 40 female rats from the F_1 generation of each of the treatment groups in the multigeneration reproduction study (see Gumbmann & Gould, 1979c, under special studies on reproduction) were maintained on their respective diets until 20% survival was reached (the males were killed during weeks 97-99 and the females were killed during weeks 110-112). Satellite groups of 5 animals of each sex were killed on day 45 and liver samples were assayed for N-demethylase activity. Body-weight gains and food intake were recorded weekly for 28 weeks and at monthly intervals thereafter. Blood samples were taken from 8 rats/sex/group at 6, 12, and 18 months and at termination and used for haematology (RBC, total and differential white-cell counts, PCV, and haemoglobin counts). Blood samples taken at 15 months and at termination were examined for serum albumin, protein, SAP, SGOT, SGPT, ornithine carbamoyl transferase (OCT), BUN, glucose, and cholesterol. Urinalyses were performed at 6-month intervals.

Weights of the liver, spleen, kidneys, heart, testes, adrenals, thyroids, and brain of all rats killed were recorded after 15 months (5 rats/sex/group) and at termination. All rats were subject to gross and histopathological examination.

Growth rates were unaffected by chlorination, but all cake-fed animals had a slightly greater growth rate than Purina controls. Food consumption was unaffected by chlorination; rats fed Purina chow ate significantly more throughout the study. Mortality was not affected by chlorination, but survival of cake-fed animals was poor and survival of the Purina controls was greatest.

No treatment-related effects of hepatic N-demethylase activity were noted in the satellite groups at day 45.

Haematological examinations at termination showed reduced PCV, haemoglobin, and RBC counts in males, especially at the 2 higher levels of treatment, while females remained unaffected. No differences due to chlorination were seen in any of the serum analyses, but SGOT, SGPT, and OCT levels were lower in cake-fed animals than in Purina controls. No treatment-related changes were observed in urinalyses, but the urines of all cake-fed rats were slightly more acidic than those of the Purina controls.

At autopsy no effects attributable to chlorination were observed on organ weights, but the absolute and relative kidney weights of rats fed cake-based diets were increased compared with Purina con-trols. The main pathological changes that were observed occurred in the stomach and kidney. Ulceration of the glandular and non-glandular stomach was seen in all females fed cake-based diets and males of the high-treatment group. There was a high incidence of nephropathy in all cake-fed animals, and mineralization of the cortico-medullary junction was noted to a greater extent in females than males. The authors con-cluded that the renal disease was the cause of the high mortality.

Although overall tumour incidence was not increased by chlorination, incidences of pituitary chromophobe adenomas and mammary fibroadenomas were higher in females receiving chlorinated cake-based diets than in cake-fed controls. However, the incidences were not higher than in Purina controls nor were they treatment related, and the significance of these changes is doubtful (Gumbmann & Gould, 1979b).

Observations in man
No information available.

Comments

In long-term and reproduction studies in which rats and mice were fed diets containing 75-79% dried cakes made from flour chlorinated at levels up to 2500 ppm, no carcinogenic, teratogenic, or other toxic effects attributable to chlorination were observed. The glomerulonephrosis and renal calcification seen in the long-term

studies in rats were considered to be due to nutritional imbalance and
did not represent a toxic response to the chlorinated cake flour.

EVALUATION
Acceptable level of treatment of flours for cake manufacturing
0-2500 ppm Cl_2.

REFERENCES

Arnold, A. (1949). Effect on dogs of flours treated with various improving agents. Cereal Chem., 26, 46-51.

Bentley, H.R., Booth, R.G., Greer, E.N., Heathcote, J.G., Hutchinson, J.B., & Moran, T. (1948). Action of nitrogen trichloride on proteins: Production of toxic derivative. Nature, 161, 126-127.

Coppock, J.B.M., Daniels, N.W.R., & Eggitt, P.W. (1960). Essential fatty acid retention in flour treatment. Chem. Ind., 1960, 17-18.

Cunningham, H.M. & Lawrence, G.A. (1976). A comparison of the distribution and elimination of oleic and chlorinated oleic acids and their metabolites in rats. Fd. Cosmet. Toxicol., 14, 283-288.

Cunningham, H.M. & Lawrence, G.A. (1977a). Absorption and metabolism of chlorinated fatty acids and triglycerides in rats. Fd. Cosmet. Toxicol., 15, 101-103.

Cunningham, H.M. & Lawrence, G.A. (1977b). Absorption and distribution studies on chlorinated oleic acid and extracts of chlorinated lipid and protein fractions of flour in rats. Fd. Cosmet. Toxicol., 15, 105-108.

Cunningham, H.M. & Lawrence, G.A. (1977c). Placental and mammary transfer of chlorinated fatty acids in rats. Fd. Cosmet. Toxicol., 15, 183-186.

Cunningham, H.M., Lawrence, G.A., & Tryphonas, L. (1977). Toxic effects of chlorinated cake flour in rats. J. Toxicol. Environ. Hlth., 2, 1161-1171.

Cunningham, H.M., & Lawrence, G.A. (1978). Effect of chlorinated lipid and protein fractions of cake flour on growth rate and organ weight of rats. Bull. Environ. Contam. Toxicol., 19, 73-79.

Daniels, D.G.H. (1960). Changes in the lipides of flour induced by treatment with chlorine dioxide or chlorine, and on storage. J. Sci. Food Agric., 11, 664-670.

Daniels, N.W.R., Frape, D.L., Eggitt, P.W., & Coppock, J.B.M. (1963). Lipids of Flour. II. Chemical and toxicological studies on the lipid of chlorine-treated cake flour. J. Sci. Food Agric., 14, 883-893.

Fisher, N., Hutchinson, J.B., Berry, R., Hardy, J., & Ginocchio, A. (1979). FMBRA Report of toxicological trials of cake made from flour treated with chlorine. Collaborative studies in rats and mice. Unpublished report from Flour Milling and Baking Research Association, Chorleywood, England.

FMBRA (1968). Submission to the Food Additives and Contaminants Committee, U.K. from Flour Milling and Baking Research Association, Chorleywood, England

Gilles, K.A., Kaelbe, E.F., & Young, V.L. (1964). X-ray spectographic analysis of chlorine in bleached flour and its fractions. Cereal Chem., 41, 412-424.

Ginocchio, A.V., Fisher, N., Hutchinson, J.B., Berry, R., & Hardy, J. (1983). Long-term toxicity and carcinogenicity studies of cake made from chlorinated flour. 2. Studies in mice, Fd. Chem. Toxicol., 21, 435-439.

Gumbman, M.R. & Gould, D.H. (1979a). Long-term feeding studies for safety evaluation of chlorine-treated flour. II. Life-span studies in mice. Unpublished report from the US Department of Agriculture.

Gumbman, M.R. & Gould, D.H. (1979b). Long-term feeding studies for the safety evaluation of chlorine-treated flour. III. Life-span studies in rats. Unpublished report from the US Department of Agriculture.

Gumbman, M.R. & Gould, D.H. (1979c). Long-term feeding studies for the safety evaluation of chlorine-treated flour. IV. Three generation reproduction and teratology studies in rats. Unpublished report from the US Department of Agriculture.

Gumbman, M.R. & Gould, D.H. (1979d). Long-term feeding studies for the safety evaluation of chlorine-treated flour. V. Six-month study in dogs. Unpublished report from the US Department of Agriculture.

Newell, G.W., Erickson, T.C., Gilson, W.E., Gershoff, S.N., & Elvejhem, C.A. (1947). Role of "agenized" flour in the production of running fits. J. Amer. Med. Assoc., 135, 760-763.

Radomski, J.L., Woodard, G., & Lehman, A.J. (1948). The toxicity of flours treated with various "improving" agents. J. Nutr., 36, 15-25.

Sollars, W.F. (1961). Chloride content of cake flours and flour fractions. Cereal Chem., 38, 487-500.

FOOD COLOURS

BROWN FK

EXPLANATION

Brown FK is prepared by coupling diazotised sulfanilic acid with a mixture of m-phenylenediamine and tolylene-2,4-diamine. The product contains 6 major coloured components, viz.:

I   1,3-diamino-4-(4'-sulfophenylazo)-benzene
II   2,4-diamino-5-(4'-sulfophenylazo)-toluene
III   1,3-diamino-4,6-bis(4'-sulfophenylazo)-benzene
IV   1,3-diamino-2,4-bis(4'-sulfophenylazo)-benzene
V   2,4-diamino-3,5-bis(4'-sulfophenylazo)-toluene
VI   1,3-diamino-2,4,6-tris(4'-sulfophenylazo)-benzene

Brown FK was evaluated at the twenty-first meeting of the Committee (Annex 1, reference 44), at which time it was noted that, in long-term studies in mice, Brown FK produced hepatic nodules and tissue pigmentation. Some of the metabolites are cardiotoxic. The reproduction/teratogenicity studies that had been performed were inadequate and no ADI could be established. A toxicology monograph was prepared.

Since the previous evaluation, additional data have become available and are summarized and discussed in the following monograph. The previously-published monograph has been expanded and is reproduced in its entirety below.

BIOLOGICAL DATA

Biochemical aspects

Absorption, distribution, excretion, and bio-transformation

After an i.p. dose to a rat of 1.5 g/kg b.w. the extremities became orange in 60 minutes, and the animal sluggish. After 24 hours the animal was normal but the urine was deep orange-yellow (Goldblatt & Frodsham, 1952).

On incubation with the contents of rat ileum and caecum, Brown FK and its coloured components underwent azo-reductive fission with formation of sulfanilic acid, a phenazine-like material (P), and ill-defined products that could be separated chromatographically. Brown FK also underwent azo-reductive fission when incubated with rat-liver homogenate, but P was not detected among the products. Oral administration of Brown FK to rats, guinea-pigs, rabbits, and pigs resulted in the excretion of sulfanilic acid in urine and faeces; P was detectable in trace amounts in the faeces, but was mainly present in caecal contents, predominantly during the first 6 hours after dosing. A "blue material" was excreted in urine. On i.p. administration to rats, Brown FK initially gave rise to brown colouring in bile; later, sulfanilic acid and the "blue material" appeared in the urine. P was not found in faeces or in caecal contents (Fore & Walker, 1967; Fore, et al., 1967).

As formed in vitro from Brown FK, P was found to consist of 2 main components, P_1 and P_2, which were identified as 1,4,7-triaminophenazine and 8-methyl-1,4,7-triaminophenazine, respectively. The "blue material" was tentatively identified as an indamine, which is an intermediate in the formation of P from the amines produced by azo-reduction of the monoazo components of Brown FK. Since 1,2,4-triaminobenzene oxidizes spontaneously in air to give the indamine and P_1, it is possible that the "blue material" and P arose from aerial oxidation of the amines formed by azo-reduction of components of Brown FK; this oxidation may have occurred during the extraction and separation of caecal contents and faeces, or in the urine after excretion (Fore et al., 1967; Walker, 1968).

In view of the complexity of Brown FK and the final mixture of metabolites, investigations have been conducted on individual components. The metabolism of 1,3-diamino-4-(4'-sulfophenylazo)-benzene and 2,4-diamino-5-(4'-sulfophenylazo)-toluene were found to be qualitatively similar, a proportion being excreted unchanged, but the bulk reductively cleaved to sulfanilic acid and the corresponding amine (the latter being acetylated before excretion). The authors expected that the metabolism of other Brown FK components would not be fundamentally different, and that the primary metabolic reactions would be products of cleavage of the azo linkages (Howes, 1969; Munday & Kirby, 1969).

Incubation of 4 individual components of Brown FK (2 mono-, 1 bis-, and 1 tris-azo-component) with rat caecal contents confirmed that azo-reduction occurred in all cases (Walker, 1968).

A summary of the products of azo reduction of the 6 major components of Brown FK is shown below:

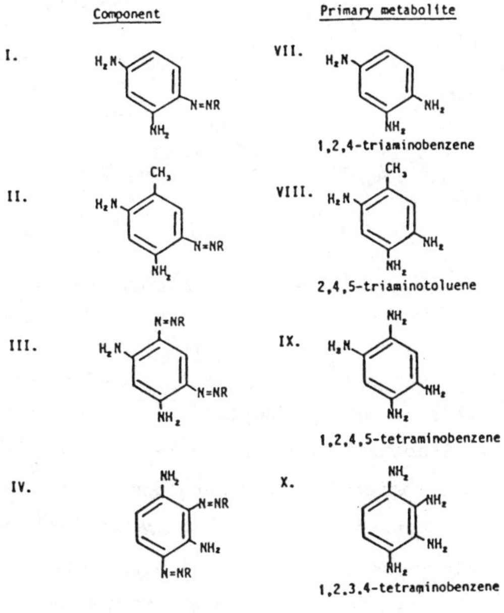

Component | Primary metabolite

I. [structure: H_2N—ring—$N=NR$, NH_2]
VII. [structure: H_2N—ring—NH_2, NH_2] 1,2,4-triaminobenzene

II. [structure: CH_3, H_2N—ring—$N=NR$, NH_2]
VIII. [structure: CH_3, H_2N—ring—NH_2, NH_2] 2,4,5-triaminotoluene

III. [structure: $N=NR$, H_2N—ring—$N=NR$, NH_2]
IX. [structure: NH_2, H_2N—ring—NH_2, NH_2] 1,2,4,5-tetraminobenzene

IV. [structure: NH_2, ring—$N=NR$, NH_2, $N=NR$]
X. [structure: NH_2, ring—NH_2, NH_2, NH_2] 1,2,3,4-tetraminobenzene

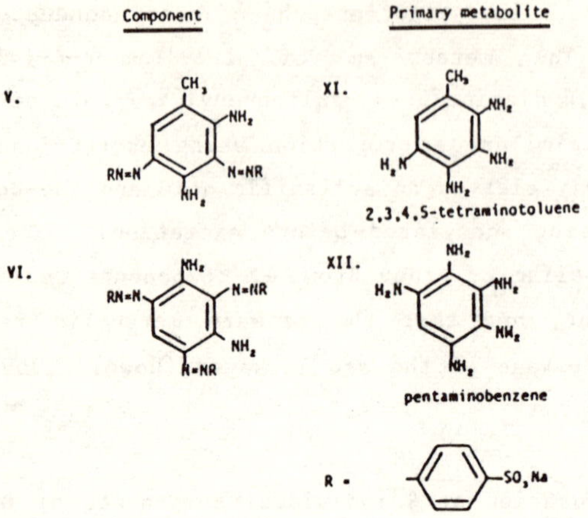

Component   Primary metabolite

V.

XI.

2,3,4,5-tetraminotoluene

VI.

XII.

pentaminobenzene

R =   —SO$_3$Na

The metabolism of 1,3-diamino-4-(4'-sulfophenylazo)-benzene (component I) in the rat is summarized below (Howes, 1969; Munday & Kirby, 1969):

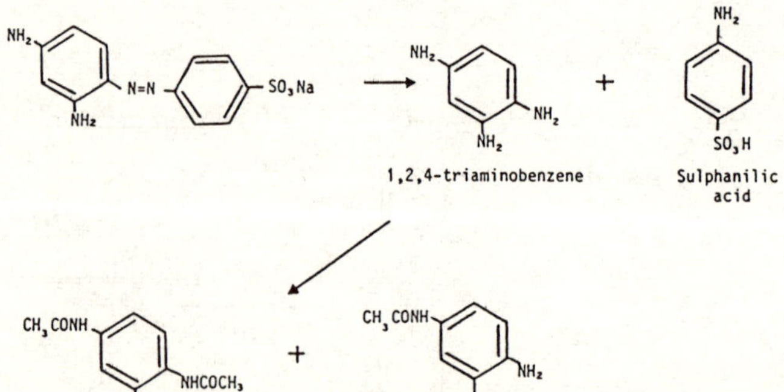

1,2,4-triaminobenzene   Sulphanilic
                                acid

1,4-diacetamido-2-aminobenzene   4-acetamide-1,2-diaminobenzene

Preliminary examination of urine from rats fed component II showed the presence of sulfanilic acid and small quantities of unchanged dye. Examination of an extract of the urine revealed the presence of 5-acetamido-2,4-diaminotoluene (the major metabolite), 2,5-diacetamido-4-aminotoluene, 2,4-diacetamido-5-amino-toluene, and 4,5-diacetamido-2-amino-toluene. Unchanged dye was identified in faecal extracts; no other dye-derived compounds were detected.

The metabolism of 2,4-diamino-5-(4'-sulfophenylazo)-toluene (component II) in the rat is summarized below (Munday, 1969):

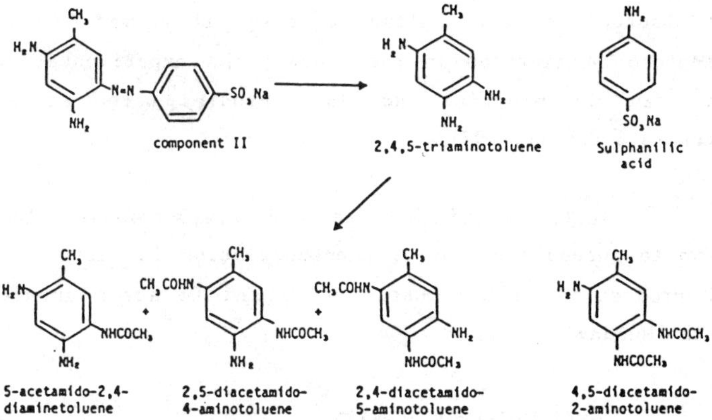

An attempt was made to determine whether 1,3-diamino-4(4'-sulfophenylazo)-benzene (component I) is reductively cleaved in humans, as in rats. Reduction of the closely-related compound, prontosil rubrum, has been shown to occur in human subjects (Fuller, 1937).

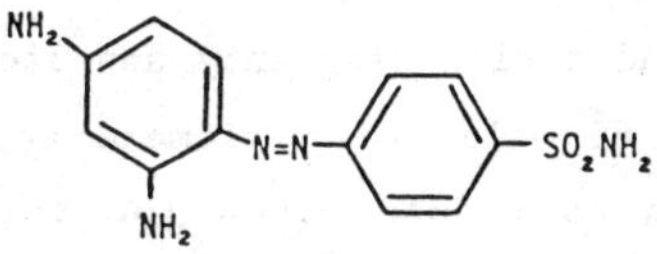

Prontosil rubrum

Administration of component I to human subjects led to no detectable unchanged dye in the urine and no appreciable urinary excretion of sulfanilic acid. It can be inferred from these results that component I is not absorbed from the intestine as such, but no information was given on the possible reduction of this compound in vivo, since it was shown that orally-administered sulfanilic acid is not absorbed in man. Sulfanilic acid, if formed from the dye, would therefore be excreted in the faeces; the experimental confirmation of this was not provided and these studies were not pursued further (Jenkins & Favell, 1971).

1,2,4-triaminobenzene and 2,4,5-triaminotoluene have been shown to uncouple oxidative phosphorylation in vitro, interfering with ATP production in the muscle cell, and to ionic imbalance with cell death (Munday, 1971).

Toxicological studies
Special studies on carcinogenicity
Rats

A carcinogenicity study on Brown FK was performed in CD rats with an in utero exposure phase. Animals of the F_0 generation (390 of each sex) were allocated to 6 groups; 2 groups were untreated (controls), 3 groups received diets containing Brown FK at constant dietary concentrations of 160, 530, or 2630 ppm (expressed as "coloured components of Brown FK"), and 1 group received sodium chloride at an amount equivalent to the amount of sodium salts received by the highest Brown FK-dose group. The animals received these diets for 14 days prior to pairing and during pregnancy and lactation. At weaning (28-31 days post-partum), 60 male and 60 female animals of the F_1 generation were selected from each dose group for the long-term study. Thereafter, the Brown FK concentration in the diet of the treated groups was adjusted to maintain constant dosages of Brown FK of 15, 50, or 250 mg/kg b.w./day (expressed in terms of coloured components); 2 groups served as untreated controls and one group (salt control) received sodium chloride equivalent to the amount of sodium salts received by the highest Brown FK-dose group. The study was terminated when mortality in any treatment group exceeded 75%, males and females being

considered separately. Accordingly, terminal sacrifices were initiated
105 weeks and 110 weeks after weaning of males and females,
respectively.

During the carcinogenicity study, animals were inspected
twice daily and palpated once weekly. Moribund animals were sacrificed
and a complete necropsy was performed on these rats and on those which
died during the course of the study. Ophthalmoscopic examinations were
performed on 20 males and 20 females from 1 control group and the top-
dose group after 26, 52, 77, and 102 weeks; the same animals were
examined on all 4 occasions and animals that died or were killed were
not replaced. Urinalysis was carried out on 10 rats of each sex from
each dose group after 103 weeks (males) or 105 weeks (females). Haematological examinations and clinical chemistry investigations were
performed on 10 males and 10 females of each group after 104 weeks
(males) or 106 weeks (females).

Rats killed in extremis or at termination were subjected to
a complete necropsy and the following organ weights were recorded:
adrenals, brain, heart, kidneys, liver, ovaries/testes, pituitary,
spleen, and thyroid. Histopathological examinations were performed on
rats of both sexes from 1 control group, the top-dose group, the salt-
control group, and on any organs from the other groups that displayed
gross abnormalities. The tissues examined histopathologically
included; adrenals, aortic arch, bone, bone marrow, brain (3 levels),
caecum, colon, diaphragm, duodenum, epididymides, eyes, heart, ileum,
jejunum, kidneys, liver (2 lobes), lungs, lymph nodes (cervical and
mesenteric), mammary gland, nasal cavities, oesophagus, ovaries,
pancreas, parathyroids, pituitary, prostate, salivary gland, sciatic
nerve, seminal vesicles, skeletal muscle, skin, spinal cord (2 levels),
spleen, stomach, testes, thymus, thyroid, tongue, trachea, urinary
bladder, and uterus (including cervix).

In the reproductive phase of the study, there were no
treatment-related effects on general condition, food intake, body-
weight gain, mating performance, conception rate, or length of
gestation. Litter size, growth, and viability of the offspring were
unaffected by treatment with Brown FK. In the carcinogenicity phase,
body-weight gains in rats of either sex receiving Brown FK at a dose of
250 mg/kg b.w./day were lower than body-weight gains in the combined

control groups; at termination the weight decrements were 14 and 10% for males and females, respectively. Food consumption was not affected by treatment, but water intakes of rats receiving 250 mg/kg b.w./day of Brown FK and of rats in the salt-control groups were greater than of untreated controls. No treatment-related effects were seen in ophthalmic or haematological examinations, nor in urine composition. Clinical chemistry investigations revealed higher creatine phosphokinase, lactate dehydrogenase, and hydroxybutyrate dehydrogenase activities in female rats of the top-dose group compared with untreated controls, but not compared with salt controls. Isocitrate dehydrogenase activities were higher among rats of both sexes receiving the highest dose of Brown FK than among controls; no effects were seen at the lower doses. In males, but not females, of the highest-dose group, plasma albumin and T4 concentrations were significantly elevated.

A total of 252 male and 236 female animals died or were killed <u>in extremis</u> during the treatment period, but the mortality distribution was unrelated to treatment. A total of 277 males and 308 females had palpable swellings during the treatment period but the distribution, frequency, and time of onset were unaffected by treatment.

No treatment-related differences were observed in absolute or relative organ weights at necropsy, but the incidence of dark thyroid glands among animals of the top-dose group was higher than controls. On histopathological examination, rats exposed to the highest-dose level of Brown FK exhibited deposition of brown pigment at particular sites (heart, skeletal muscle, diaphragm, tongue, thyroid, caecum, and hepatic kupffer cells) but this was not associated with any tissue reaction and was not observed at dose levels of 15 or 50 mg/kg b.w./day of Brown FK. Treatment with Brown FK was not associated with enhancement of neoplasia at any site (there was a reduced incidence of tumours in animals of the top-dose group compared with controls).

Chronic myocarditis occurred with high frequency in all groups, but the incidence and severity were no greater in treated animals than in controls. In females, but not males, exposure to the highest-dose level of Brown FK was associated with an increased incidence of pelvic nephrocalcinosis; this effect was not seen at lower-dose levels. There was an increase in cystic distension of the follicles of the thyroid in high-dose group females, where the lesion was seen in 40.7% of the animals, compared with 17.2% of the controls; there was no evidence of any effect at lower-dose levels.

The authors concluded that Brown FK was not carcinogenic in rats under the conditions of the experiment and that the no-effect level was 50 mg/kg b.w./day (Tesh et al., 1980; Amyes et al., 1983; Roe, 1983).

Special studies on mutagenicity

Brown FK and its constituents were assayed for mutagenicity in Salmonella typhimurium TA1535, TA1537, and TA1538 when activated by a rat-liver supernatant fraction. Mutagenicity was linearly dose-dependent in the range 0–3 mg/plate, with activities ranging from 22 to 50 times the spontaneous mutation frequency. One sample of Brown FK was mutagenic in the absence of metabolic activation, producing a 16-fold increase in mutation at 4 mg/plate. Two major constituents of Brown FK, 2,4-diamino-5-(4'-sulfophenylazo)toluene (II) and 1,3-diamino-4-(4'-sulfophenylazo)benzene (I), each present at about 18% in the complete colour, were mutagenic in TA1538. Mutagenicity was linearly dose-related in the range 0–1 μmol/plate, with slopes of 1.5 mutants/nmol for compound I and 0.35 mutants/nmol for compound II. This activity was dependent on metabolic activation. Four other major constituents were inactive, as was sulfanilic acid, the major excretion product. The combined effects of compounds I and II could largely account for the mutagenicity of Brown FK (Venitt & Bushell, 1976).

Special studies on reproduction
Rats

A multigeneration reproduction study was carried out in rats in which groups of 24 of each sex were given Brown FK in the diet at a level of 300 ppm for 5 weeks post-weaning, then at 600 ppm for 3 successive generations; control animals received stock diet. After weaning, the parental animals and offspring not selected for breeding in the succeeding generation were culled. Haematological and clinical chemistry studies were performed on 10 animals of each sex from the parents and offspring, and autopsies carried out. Selected organ weights were recorded at autopsy on the parents and on 30 animals of each sex per group from the offspring. Histopathology was performed on F_3 weanlings, 10 animals of each sex per group.

Fertility, number of young per litter, birth weights, growth rates, gestation indices, viability indices, and lactation indices were unaffected by treatment; there was no indication of increased mortality in utero. No treatment-related lesions were observed at autopsy nor on histopathological examination of F_3 weanlings. In clinical observations (plasma biochemistry/haematology) and in organ weights, occasional statistically-significant differences were seen; these were not consistent and, in the absence of pathological lesions, were not considered to be of toxicological significance. Under the conditions of the study, Brown FK was without adverse effect on reproductive performance (Wilson et al., 1983b).

Special studies on teratogenicity

Rats

Groups of 35 virgin female Wistar rats were mated with virgin males and were fed diets containing 0, 0.03, 0.15, or 0.6% Brown FK from day 0 to day 19 post coitus. Additional groups received 0.6% sodium chloride (salt control) or aspirin (250 mg/kg b.w./day). Successful pregnancies were achieved in 32-35 animals per dose group. Five pregnant animals per group were allowed to litter normally and then raise their offspring to weaning. The remaining animals were sacrificed on day 21 of gestation and the foetuses removed by Caesarian section. Two-thirds of the foetuses were examined for gross soft-tissue abnormalities, then cleared and stained with Alizarin red for examination for skeletal defects. The remaining one-third were examined for soft tissue defects using Wilson's technique. No treatment-related abnormalities were observed in any of the groups receiving Brown FK, and this was confirmed in the groups which were examined at weaning. Aspirin used as a positive control induced teratological defects in foetuses and in young reared to 21 days post partum (Unilever, 1978).

Special studies on pigment deposition

After feeding Brown FK to rats and mice, pigment was found in heart, skeletal muscle, tongue, diaphragm, thyroids, brain, liver,

kidneys, spleen, lungs, pancreas, bladder, testes, ovary, uterus, skin,
stomach, duodenum, ileum, brown fat, and bone marrow. In addition, a
pigment has been detected in the plasma of rats.

Staining tests commonly used to identify lipofuscin were
negative with the exception of the test for metachromasia with tolui-
dine blue. The tests applied were as follows:

	Usual response of known lipofuscin	Response of Brown FK-induced pigment
Test for iron	negative	negative
Sudan fat stains	positive	negative
Reduction of ferric salts	positive	negative
Reduction of ammoniacal silver salts	positive	negative
Basophilic properties	positive	negative
Periodic acid - schiff reaction	positive	negative
Acid fastness	acid fast	negative
Toluidine blue at pH3	stains meta-chromatically green	greenish

Two further histochemical tests clearly differentiated
between lipofuscin and the Brown FK-induced pigment:
-   Potassium permanganate/oxalic acid bleached lipofuscin, but
    not the Brown FK-induced pigment.
-   Sodium dithionite bleached both lipofuscin and the
    Brown FK-induced pigment. However, after rinsing and
    allowing to stand in air, the Brown FK-induced pigment
    reappeared; lipofuscin was permanently bleached.

The Brown FK-induced pigment does not fluoresce in
ultra-violet light. On the other hand, all samples of lipofuscin which

have been examined were fluorescent. Pigment has been found in the thyroid, brown fat, and bone marrow; these tissues have not been recorded as being sites for lipofuscin deposition. Furthermore, a coloured substance has been demonstrated in the plasma, which has never been found with lipofuscin.

The speed at which the Brown FK-induced pigment is deposited is uncharacteristic of lipofuscin information. In acute studies, pigment has been seen in the intestinal wall and villi within 24 hours of feeding the dye, and in the kidney after 5 days. Pigment masses produced in macrophages either _in vivo_ after the intraperitoneal injection of Brown FK into mice, or _in vitro_, when Brown FK was incorporated in the macrophage culture medium, appeared identical. Tests for lipofuscin proved negative; the pigment in the macrophages closely resembled that seen in macrophages in stained sections of tissues from rats and mice fed Brown FK.

Electron microscope studies have identified differences in morphology between lipofuscin and the Brown FK-induced pigment. In aged rats and mice fed Brown FK, conjugate forms were observed in which induced pigment and control lysosomal material appeared in the same membrane-limited body (Hope, 1971).

It is likely that a component of Brown FK is oxidized within the cell to 1,4,7-triaminophenazine:

a brown water-insoluble material.

This would explain the behaviour of the pigment with sodium dithionite, which reduces the phenazine ring to the 5,10-dihydro derivative, which is probably colourless. After exposure to air re-oxidation would occur. Thus, the pigment may not represent evidence of sub-lethal cell damage, but is, instead, an insoluble oxidation product of a dye metabolite.

1,2,4-triaminobenzene was very rapidly oxidized to 1,4,7-triaminophenazine by a mitochondrial suspension; no phenazine deriv-atives were detected with triaminotoluene under the same circumstances (Kirby, 1968b).

1,4,7-triaminophenazine is a brown, water-insoluble, mater-ial which is very readily formed from 1,2,4-triaminobenzene (Muller, 1889).

Special studies on pigment in tissues

Histological tests were applied to sections of hearts and livers from female rats fed Brown FK at doses of 0, 15, 50, or 250 mg/kg b.w./day for 106-108 weeks. These tissues were obtained from animals used in the carcinogenicity study (see above). In addition, several tissues taken from weanling rats discarded at the end of the in utero phase were screened to determine if any pigment was present at the start of the long-term study due to transplacental transfer of Brown FK or through exposure during lactation/creep feeding.

Rats fed Brown FK for over 2 years at a dose-level of 250 mg/kg b.w./day displayed substantial pigment deposition in the heart and liver, but no effects were seen in these organs at the lower dose-levels. Brown FK-associated pigment was not found in tissues from weanling rats exposed to Brown FK in utero. The no-effect level for pigment accumulation in the long-term study was 50 mg/kg b.w./day. Differential tests showed that the pigment was not formalin pigment or haemosiderin, but tests aimed at differentiating between lipofuscin and Brown FK-induced pigment gave inconclusive results (Wilson et al., 1983a).

Special studies on components I and II of Brown FK
Mice

Groups of 3 male and 3 female C57B1 mice were fed diets containing 0, 0.5, or 1.0% of the "azobenzene" or "azotoluene" components of Brown FK (components I & II, respectively) for 6 weeks. With both components, the thyroids were dark and the intestines and squamous portions of the stomach were stained salmon-pink. Heart lesions were seen in all mice fed 1% component II, but not in those given component I. More pigment was seen in mice fed component I, less in those fed component II (Kirby, 1968a).

Rats

Groups of 3 male and 3 female Colworth-Wistar rats were fed component I or component II at dietary levels of 0, 0.5, or 1% for 6 weeks. The thyroids of rats receiving component I were dark and the hearts, muscles, and brains were stained. However, the intestines were stained only slightly. Pale hearts and meningeal haemorrhage were seen with component II, otherwise pigmentation was as with component I. One-sixth of the rats treated with component I and four-fifths of the rats treated with component II had heart lesions. More pigment was seen histologically in component I-treated rats, less in component II-treated animals (Kirkby, 1968a).

Special studies on amine metabolites of Brown FK

Amines derived from Brown FK and from its 2 myotoxic components, components I and II, were injected i.v. into rats in single doses of 3.13-25 mg/kg. The mixture of amines from Brown FK was also injected into mice in the same range of doses. Cardiac and muscular lesions were produced by the amines in both species. These amines are biological degradation products in the intestine. The finding that orally-administered Brown FK is myotoxic in rats but not in mice is probably due to differences in the intestinal flora in the 2 species (Walker et al., 1970).

In another study, 1,2,4-triaminobenzene was given to groups of 6-7 rats orally 5 days/week for 2 weeks at 50, 60, 75, or 100 mg/kg b.w./day. Six rats out of 7 receiving 100 mg/kg/day died after 3 doses with severe heart lesions; 7/11 on 75 mg/kg/day also died after 3 doses. Heart pigmentation occurred after 5 days' treatment or longer. Animals on lower doses showed both extensive heart pigmentation and cardiac necrosis (Mulky et al., 1969).

1,2,4,5-Tetraaminobenzene was given to groups of 6 rats orally 5 days/week for 2 weeks at 150 or 200 mg/kg b.w./day. 1,2,3,4-Tetraaminobenzene was given to groups of 6 rats orally 5 days/week for 2 weeks at 125 or 166 mg/kg b.w./day. No frank heart lesions and only instances of diffuse increase in interstitial cells in the heart were observed. No heart or thyroid pigment deposition was observed (Mulky et al., 1969).

Acute toxicity

Species	Route	LD_{50} (mg/kg b.w.)	Reference
Mouse	oral	>2,000 (with salt)	Grasso et al., 1968a
	oral	1,100-2,250 (with salt)	Edwards & Wilson, 1966
	oral	960-1,149 (no salt)	Edwards & Wilson, 1966
	i.p.	1,500-2,000 (with salt)	Grasso et al., 1968a
	i.p.	960-1,720 (with salt)	Edwards & Wilson, 1966
	i.p.	840-880 (no salt)	Edwards & Wilson, 1966
Rat	oral	>8,000 (with salt)	Grasso et al., 1968a
	oral	900-1,910 (with salt)	Edwards & Wilson, 1966
	oral	780-970 (no salt)	Edwards & Wilson, 1966
	i.p.	750-1,150 (with salt)	Grasso et al., 1968a
	i.p.	1,100-2,250 (with salt)	Edwards & Wilson, 1966
	i.p.	960-1,150 (no salt)	Edwards & Wilson, 1966
Guinea-pig	oral	3,000 (with salt)	Edwards & Wilson, 1966
	oral	2,610 (no salt)	Edwards & Wilson, 1966
	i.p.	900 (with salt)	Edwards & Wilson, 1966
	i.p.	780 (with salt)	Edwards & Wilson, 1966
Rabbit	oral	450-680 (with salt)	Edwards & Wilson, 1966
	oral	390-590 (no salt)	Edwards & Wilson, 1966
Chicken	oral	>10,000 (with salt)	Edwards & Wilson, 1966
	oral	>8,700 (no salt)	Edwards & Wilson, 1966

For all species, animals dying did so from within a few minutes to 96 hours. Many animals, after either oral or i.p. treatment, showed lack of coordination, hypersensitivity, and hyperactivity; convulsions usually preceded death (Edwards & Wilson, 1966).

Meningeal congestion or haemorrhage was seen at post-mortem examination in rats and mice which died following both oral and i.p. treatment with 3.4 g/kg Brown FK as a 10% solution. This was the

highest dose administered, and the condition may have been present to a
lesser degree in animals treated with lower levels of Brown FK, but
post-mortem identification of the lesion was made difficult by tissue
colouration. The meningeal congestion/haemorrhage was probably caused
by the sodium chloride in the dye solution, since this lesion is
observed after administration of hypertonic solutions of sodium
chloride to rats. In this instance, the lowest dose levels at which
the meningeal lesion was observed were 6.0 g/kg of sodium chloride
orally as a 10% solution and 4.0 g/kg i.p. as a 5% solution (Edwards &
Wilson, 1966).

Following oral intubation, external tissue colouration was
apparent after several hours in rats and guinea-pigs. No colouration
of the tissues was seen in rabbits and chickens. After i.p. injection,
external tissue colouration was apparent and intense after a few
minutes in rats and guinea-pigs. Colour was seen in the faeces of
rats, mice, rabbits, and guinea-pigs up to 24 hours after oral
treatment; it was also excreted in the urine of rats, mice, guinea-
pigs, and rabbits within 15 minutes of either oral or i.p. treatment
(Edwards & Wilson, 1966).

Hearts from some rats and mice surviving for 21 days after
treatment were examined histologically. Degenerative lesions were
found in 15% of rats given orally 1-2.5 g/kg b.w. Brown FK, but not in
rats given 3.37 g/kg b.w. Brown FK. The same lesions were found in 50%
of mice given orally 0.9 g/kg b.w. Brown FK, but not when given 0.6,
1.35, or 2.03 g/kg b.w. Brown FK. When given i.p., 25-60% of mice
showed lesions at 0.75 and 1.03 g/kg b.w. Brown FK (Edwards & Wilson,
1966).

Short-term studies
Mice

Groups of 10 male or 10 female mice received the colour
(either fresh or stored) at a level of 1 g/kg daily for 3 weeks. A
significant reduction in body-weight gain was noted in the mice
receiving the stored solution, but not in those receiving fresh

solution. One male and 1 female receiving the fresh solution showed cardiac lesions (BIBRA, 1964).

Daily oral or i.p. doses up to 2 g/kg or 1 g/kg, respectively, for 43 days to groups of 10 or 12 mice were well-tolerated (Grasso et al., 1968a).

Groups of 10 male and 10 female mice (Colworth C57Bl strain, initially 6 weeks old) were fed for 90 days on a synthetic diet containing 0, 0.05, 0.075, 0.10, 0.25, 0.50, 0.75, 1.0, or 2.0% Brown FK (equivalent to 0, 0.025, 0.0375, 0.05, 0.125, 0.25, 0.375, 0.50, or 1.0% Brown FK-coloured components) containing 51% dye component and 47% salt. A further group of 20 mice were fed synthetic diet containing added 1.0% sodium chloride as a control for the additional dietary salt derived from the Brown FK. At the 0.125% dietary colour level, pigment deposition occurred in tissues. At 0.25% and above there was splenic enlargement, at 0.50% the liver and heart were enlarged, and at 1.0% there was reduced growth, poor food utilization, liver, spleen, heart, and testicular enlargement, and histological evidence of degenerative heart lesions. The thyroids, muscle, intestine and squamous part of the stomach were pigmented (Ashmole et al., 1958).

Rats

Groups of animals received the colour at a level of 0.5 g Brown FK per kg b.w. for 3 weeks, orally or i.p. Twenty rats were dosed orally, of which 6 animals died after having been administered between 5 and 11 doses. Post-mortem examination of rats dying during the test or killed at the end revealed general tissue-staining in 5 rats. Of 18 hearts examined histologically, 8 showed degenerative lesions, and a brown pigment was observed in small amounts in 9 hearts after 3 weeks.

In the multiple-dose i.p. test, 8 rats were treated and none died during the treatment period. General organ-staining was observed in all animals at post-mortem examination. Hearts from 7 rats were examined microscopically and degenerative lesions were found in 1 heart and small amounts of pigment in 3 hearts after 3 weeks (Kirkby, 1968a).

No ill-effects were seen in 3 weanling rats given a 0.1% solution Brown FK for 28 days, the intake being equivalent to 15 mg/day (Goldblatt & Frodsham, 1952).

Administration of 2 or 3 oral doses of 1 g Brown FK/kg b.w. to rats induced a myopathy in cardiac and skeletal muscles character- ized by multiple vacuoles about 1-2 micrometers in diameter. Ultra- structurally, these were shown to consist of areas of fibrollolysis. Histochemically, the myopathy was accompanied by a moderate increase in acid phosphatase activity and by a loss of phosphorylase activity. Subsequently, complete lysis of the affected fibres ensued. In the heart, lysis was followed by macrophage invasion and fibroblastic proliferation, and in skeletal muscle by regeneration. The occurrence of lipofuscin in muscle fibres and in macrophages was scanty and erratic. When Brown FK was given in the diet at a level of 2%, fibrillolysis and an increase in the number and electron-density of lysosomes was observed ultrastructurally during weeks 2 to 3 of the test. These changes were accompanied by a marked elevation of histochemically-demonstrable acid phosphatase. Progressive deposition of lipofuscin was the principal pathological feature during weeks 3 to 12 (Grasso et al., 1968b).

Daily oral doses of up to 2 g/kg Brown FK for 43 days to groups of 10 rats induced rapid loss of weight and death, with severe damage to cardiac and skeletal muscle, characterized by vacuolar myopathy and lipofuscin deposition. Of 3 pure components of Brown FK studied, component II, and to a lesser extent component I, produced similar, but not identical lesions to those induced by the parent colour after repeated oral doses of 0.5 g/kg. Ultrastructural studies confirmed an extensive loss of myofibrillar elements, and histochemical studies revealed a loss in the activity of mitochondrial enzymes. Similar i.p. injections in doses up to 1.0 g/kg for 43 days to groups of 10 or 12 rats did not have any effect on the heart or skeletal muscle (Grasso et al., 1968b).

Experiments were performed using groups of 10-12 rats receiving the colour at levels of 0.1 or 1 g/kg orally or 0.1, 0.25, or

1 g/kg i.p. daily for up to a maximum of 43 doses. A specific cardiac
lesion was identified at the oral dose of 1 g/kg. There were large
areas of myocardial necrosis and replacement by large mononuclears,
with involvement of the sub-pericardial region and endocardium. Some
myocardial cells had lost their stainable cytoplasm and appeared only
as empty sheaths. When administered i.p., the colour produced little
or no cardiac damage at any dose tested. At these high doses of 1
g/kg, most animals showed congestion, fatty change, or necrosis of the
liver with hydropic degeneration of the kidney. There was no obvious
splenomegaly. Daily doses of 100 mg/kg by stomach tube produced 2
pericardial and 1 sub-pericardial lesions. In addition, early hydro-
ponic degeneration of the kidney was seen in 2 rats, with 1 of these
animals also showing fatty change in the liver (BIBRA, 1964).

Administration of Brown FK (purity 80.0%) at dietary levels
of 0, 0.001, 0.01, 0.1, or 1.0% for 150 days showed no adverse effects
on growth, food consumption, haematological indices, liver and kidney
function, or organ weights. One male rat at the 1.0% level showed
typical myocardial changes; other rats showed deposits of lipofuscin,
especially in females. The no-effect level was 0.1% (Gaunt et al.,
1968).

Groups of 12 male and 12 female rats (Colworth-Wistar
strain, initially 3-4 weeks old) were fed for 112 days on a commercial
stock diet containing 0, 0.05, 0.1, 0.5, 1.0, or 2.0% Brown FK (0,
0.025, 0.05, 0.25, 0.5, or 1.0% Brown FK-coloured components) containing 51% dye component and 47% salt. A further group of 24 rats were
fed the commercial stock diet containing added 1.0% sodium chloride as
a control for the additional dietary salt derived from the Brown FK.
In addition, to eliminate damage to the heart from cardiac puncture in
any rat kept to 16 weeks, groups of 6 male and 6 female rats were fed 6
weeks on powdered stock diet containing 0, 0.05, 0.5, or 2.0% Brown FK
(0, 0.025, 0.25, or 1.0% Brown FK-coloured components). A group of 12
rats also received 1.0% sodium chloride added to the basic diet. All
these rats were used for biochemical tests during weeks 0-6 after which
they were killed. At the 0.25% level tissue pigmentation appeared, at
0.5% liver enlargement occurred, and at the 1% level there was reduced

growth, poor food utilization, enlargement of the liver, testes, and thyroid, histological evidence of degenerative heart lesions, increased urinary indican excretion, and elevation of SGOT. The intestine, squamous portion of the stomach, and the thyroid were stained. The no-effect level was 0.05% (Ashmole et al., 1966).

Pigs

Groups of female and male pigs were given doses of 0, 100, 250, or 500 mg Brown FK/kg/day for 24 weeks without adverse effects on growth, food consumption, haematological indices, liver and kidney function, or organ weights. Lipofuscin was widely distributed in animals of both sexes at all dose levels in one or more organs. The liver was particularly affected in that lipofuscan deposition was accompanied by increased lysosomal enzyme activity, which was more marked at the higher dose-levels. It was also seen in the heart in males, where it was associated with an increased acid phosphatase activity, and in the kidneys at the highest-dose level in females and at all levels in males. A no-effect level could not be determined in this study (Gaunt et al., 1968).

Long-term studies
Mice

Groups of 40 male and 40 female Colworth C57Bl mice were fed for 80 weeks on a synthetic diet containing 0, 0.0125, 0.0375, 0.075, 0.125, or 0.625% Brown FK-coloured components (the Brown FK used in this study contained 62.5% coloured components). Only at the 0.625% level was there reduced growth and food utilization and increased mortality among females. There were increased liver, kidney, spleen, brain, and testes weights, evidence of splenic haemopoisis, and increased myocardial fibrosis. Heart weights were increased at the 0.125% level. Increased hepatic nodules were seen at 0.075% and higher levels and pigment deposition at 0.0375% and higher levels. At termination, after 80 weeks, the number of animals with nodules at the different dose levels was 26, 23, 27, 56, 42, and 64, respectively. Increased hepatic nodules were observed at 0.075% and higher levels. The number of mice with hepatocellular carcinoma were 3, 2, 0, 5, 6, and 2 at the various dose levels, respectively. Pigment deposition was observed at dose levels of 0.0375% and higher (Wilson et al., 1970).

Rats

Groups of 32 male or 36 female Colworth-Wistar rats were fed for 2 years on a synthetic diet containing 0, 0.01, 0.03, 0.06, 0.1, or 0.5% Brown FK-coloured components (Brown FK used in this study contained 54.2% coloured components). Only at the 0.5% level was there increased splenic weight and hepatic granulomata. Pigment deposition was seen at dose levels of 0.06% and higher. The no-effect level for pigment deposition was 0.03%; it was 0.06% when based on toxicity evidence (Wilson et al., 1971).

Observations in man
No information available.

Comments

The carcinogenicity study did not reveal any increase in the incidence of tumours, nor did the reproduction and teratogenicity studies show any adverse effects on reproductive function.

The myopathy seen in rats given high doses of Brown FK in short-term studies affected all striated muscle, was accompanied by pigment deposition, and was dose-dependent with a high threshold. In the long-term/carcinogenicity study in rats, the observed myocarditis was not dose-related and there was no pigmentation in animals of the low-dose group nor controls. However, deficiencies in histopathological examination of tissues from the low- and intermediate-dose groups hindered the establishment of a no-effect level in this study.

In earlier long-term studies, the no-effect level (with respect to pigment deposition) was 0.03% in the diet, equivalent to 15 mg/kg b.w./day, based on the coloured components of Brown FK.

EVALUATION
Level causing no toxicological effect
Mouse: 0.0125% in the diet, equivalent to 19 mg/kg b.w./day.
Rat: 0.03% in the diet, equivalent to 15 mg/kg b.w./day.

Estimate of temporary acceptable daily intake for man
0-0.075 mg/kg b.w. (based on colour components)

Further work or information
Required by 1986
A complete histopathological examination of tissues from the low- and intermediate-dose groups in the long-term/carcinogenicity study in rats.

REFERENCES

Amyes, S.J., McSheehy, T.W., & Whitney, J.C. (1983). Brown FK: Life-span combined toxicity and oncogenicity study in rats pre-exposed in utero. 2. Toxicity and oncogenicity phase. Unpublished report No. 82/URL012/573 from Life Science Research, Essex, U.K. Submitted to WHO by Unilever Ltd.

Ashmole, R.T., Campbell, P., Kirkby, W.W., & Wilson, R. (1966). Effects of feeding dietary Brown FK to rats for six and 16 weeks. Unpublished report from Unilever Research Laboratories. Submitted to WHO by Unilever Ltd.

Ashmole, R.T., Kirkby, W.W., & Wilson, R. (1958). Thirteen week mouse feeding trial. Unpublished report from Unilever Research Laboratories. Submitted to WHO by Unilever Ltd.

BIBRA (1964). Unpublished research report No. 5/1964 from British Industrial Biological Research Association, Carshalton, Surrey, England. Submitted to WHO.

Edwards, K.B. & Wilson, R. (1966). Acute toxicity of Brown FK in rats, mice, guinea-pigs, rabbits and chickens. Unpublished report from Unilever Research Laboratories.

Fore, H. & Walker, R. (1967). Studies on Brown FK. I. Composition and synthesis of components. Fd. Cosmet. Toxicol., 5, 1-9.

Fore, H., Walker, R., & Golberg, L. (1967). Studies on Brown FK. II. Degradative changes undergone in vitro and in vivo. Fd. Cosmet. Toxicol., 5, 459-473.

Fuller, A.T. (1937). Is p-aminobenzene sulphonamide active agent in prontosil therapy? Lancet, 1, 194-198.

Gaunt, I.F., Hall, D.E., Grasso, P., & Golberg, L. (1968). Studies on Brown FK. V. Short-term feeding studies in the rat and pig. Fd. Cosmet. Toxicol., 6, 301-312.

Goldblatt & Frodsham (1952). Private communication from ICI (unpublished report).

Grasso, P., Gaunt, I.F., Hall, D.E., Golberg, L., & Batstone, E. (1968a). Studies on Brown FK. III. Administration of high doses to rats and mice. Fd. Cosmet. Toxicol., 6, 1-11.

Grasso, P., Muir, A., Golberg, L., & Batstone, E. (1968b). Cytopathic effects of Brown FK on cardiac and skeletal muscle in the rat. Fd. Cosmet. Toxicol., 6, 13-24.

Hope, J. (1971). Ultrastructure of the pigment induced in various tissues of the rat by long-term feeding of the dye Brown FK. Unpublished report from Unilever Research Laboratories. Submitted to WHO by Unilever Ltd.

Howes, D. (1969). Metabolism of ^{14}C labelled 1,3-diamino-4-(p-sulphophenylazo)benzene, a component of the dye Brown FK, in the rat. Unpublished report from Unilever Research Laboratories. Submitted to WHO by Unilever Ltd.

Jenkins, F.P. & Favell, D.J. (1971). Metabolism of the "monoazobenzene" component of Brown FK in human subjects. Unpublished report from Unilever Research Laboratories. Submitted to WHO by Unilever Ltd.

Kirkby, W.W. (1968a). Effects of Brown FK and two of its constituents on pigment deposition and lesions in rats and mice. Unpublished report from Unilever Research Laboratories. Submitted to WHO by Unilever Ltd.

Kirkby, W.W. (1968b). Nature of the pigment induced in tissues of rats and mice fed Brown FK. Unpublished report from Unilever Research Laboratories. Submitted to WHO by Unilever Ltd.

Mulky, M.J., Mundy, R., Ashmole, R.T., & Kirkby, W.W. (1969). Evaluation of the terminal causative agent in Brown FK induced myopathy and pigment deposition. Unpublished report from Unilever Research Laboratories. Submitted to WHO by Unilever Ltd.

Muller, E. (1889). Chem. Ber., 22, 856.

Munday, R. (1969). Metabolism of 2,4-diamino-5-(ρ-sulphophenylazo) toluene. Unpublished report from Unilever Research Laboratories. Submitted to WHO by Unilever Ltd.

Munday, R. (1971). Uncoupling of oxidative phosphorylation by Brown FK metabolites. Unpublished report from Unilever Research Laboratories. Submitted to WHO by Unilever Ltd.

Munday, R. & Kirkby, W.W. (1969). Metabolism of 1,3-diamino-4-(ρ-sulphophenylazo)benzene. Unpublished report from Unilever Research Laboratories. Submitted to WHO by Unilever Ltd.

Roe, F.J.C. (1983). Histopathological evaluation of sections derived from URL/12/BFK lifespan combined toxicity and oncogenicity study in rats pre-exposed in utero. Unpublished report from Unilever Research Laboratories. Submitted to WHO by Unilever Ltd.

Tesh, J.M., McSheehy, T.W., McAnulty, P.A., & Collier, M.J. (1980). Brown FK: Lifespan combined toxicity and oncongenicity study in rats pre-exposed in utero. 1. Reproductive phase. Unpublished report No. 80/URL012/051 from Life Science Research, Essex, U.K. Submitted to WHO by Unilever Ltd.

Unilever (1978). Teratology study on Brown FK in the Colworth-Wistar rat. Unpublished report from Unilever Research Laboratories. Submitted to WHO by Unilever Ltd.

Venitt, S. & Bushell, C.T. (1976). Mutagenicity of the food colour Brown FK and constituents in Salmonella typhimurium. Mutation Research, 40, 309-316.

Walker, R. (1968). Intestinal degradation of azo food colours with particular reference to Brown FK. Ph.D. Thesis, University of Reading, U.K.

Walker, R., Grasso, P., & Gaunt, I.F. (1970). Myotoxicity of amine metabolites from Brown FK. Fd. Cosmet Toxicol., 8, 539-542.

Wilson, R., Gellatly, J.B.M., Kirkby, W.W., & Ashmole, R.T. (1970). Biological evaluation of Brown FK: 80-week mouse feeding trial. Unpublished report from Unilever Research Laboratories. Submitted to WHO by Unilever Ltd.

Wilson, R., Gellatly, J.B.M., Kirkby, W.W., & Ashmole, R.T. (1971). Biological evaluation of Brown FK: 2-year rat feeding trial. Unpublished report from Unilever Research Laboratories. Submitted to WHO by Unilever Ltd.

Wilson, R., Hague, P.H., & Hardy, W.S. (1983a). Brown FK: Lifespan combined toxicity and oncogenicity study in rats pre-exposed in utero: Examination of selected tissues for Brown FK associated pigment. Unpublished report from Unilever Research Laboratories. Submitted to WHO by Unilever Ltd.

Wilson, R., McCormick, S.G., Cook, H.J., Norris, L., Robinson, J.A., & Williams, T.C. (1983b). Evaluation of the effects of food colour Brown FK on the fertility and reproductive performance of rats. Unpublished report from Unilever Research Laboratories. Submitted to WHO by Unilever Ltd.

CARAMEL COLOURS

EXPLANATION

In the period since 1972, when the fifteenth report of the Committee was published (Annex 1, reference 26), caramel colours have been classified into 4 classes which differ in their method of manufacture, composition, functional properties, and application.

1. **Caramel Colour I** (synonyms: plain caramel, caustic caramel, and spirit caramel); this class is prepared by the controlled heat treatment of carbohydrates with alkali or acid.

2. **Caramel Colour II** (synonyms: caustic sulfite process caramel); this class is prepared by the controlled heat treatment of carbohydrates with sulfite-containing compounds.

3. **Caramel Colour III** (synonyms: ammonia caramel, ammonia process caramel, closed-pan ammonia process caramel, open-pan ammonia process caramel, bakers' caramel, confectioners' caramel, and beer caramel); this class is prepared by the controlled heat treatment of carbohydrates with ammonium compounds.

4. **Caramel Colour IV** (synonyms: ammonia sulfite process caramel, sulfite ammonia caramel, sulfite ammonia process caramel, acid-proof caramel, beverage caramel, and soft-drink caramel); this class is prepared by the controlled heat treatment of carbohydrates with ammonium-containing and sulfite-containing compounds.

Caramel colours were reviewed at the eighth, thirteenth, fifteenth, sixteenth, eighteenth, twenty-first, and twenty-fourth meetings of the Committee (Annex 1, references 8, 19, 26, 30, 35, 44, & 53). The thirteenth meeting concluded that a toxicological distinction between caramels produced commercially and caramel formed in cooked foods or when sucrose is heated is unwarranted except with caramel prepared by processes using ammonia or ammonium salts. This conclusion was endorsed by the fifteenth meeting and an ADI "not limited" was allocated to caramel colours prepared by processes other than those involving ammonia or ammonium salts; a temporary ADI of 0-100 mg/kg b.w. was allocated to caramel colours produced by the ammonia process. The sixteenth meeting reviewed further information on the composition of caramel colours produced by the ammonia process, including the 4-methylimidazole content. Revised specifications were prepared and the previously-established temporary ADI was maintained. The specifications for this class were further revised by the eighteenth meeting and the temporary ADI was extended pending the results of long-term and reproduction studies on caramel colours prepared by the ammonia or ammonia sulfite process.

The twenty-first meeting of the Committee noted that the specifications for caramel colour (ammonia process) were ambiguous, since they appeared to cover caramel colours manufactured by the ammonia sulfite process as well. Separate specifications were prepared for caramel colour (ammonia process) and caramel colour (ammonia sulfite process), which have since been designated caramel colours III and IV, respectively. The two classes have been shown to differ in toxicity. The principal toxic effect of caramel colour III is depression of circulating lymphocytes and leucocytes, and a no-effect level could not be determined; accordingly, the temporary ADI was revoked. The temporary ADI of 0-100 mg/kg b.w. was retained for caramel colour IV pending the submission of reports of adequate carcinogenicity/teratogenicity studies.

The twenty-fourth meeting extended the temporary ADI for caramel colour IV and confirmed that caramel colour II had no ADI since it was not included in the ADI for caramel colour I nor the temporary ADI for caramel colour IV.

Since the previous evaluations, additional data have become available and are summarized and discussed in the following monographs. Previously-published monographs have been expanded and are reproduced in their entirety under the class of caramel colour to which they relate.

CARAMEL COLOUR I

EXPLANATION

At the thirteenth and fifteenth meetings (Annex 1, references 19 & 26), the Committee concluded that caramel colour I is a natural constitutent of the diet and is acceptable as an additive. An ADI "not limited" was allocated at the fifteenth meeting.

BIOLOGICAL DATA
Biochemical aspects
No information available.

Toxicological studies
Special study on mutagenicity
Two samples of caramel colour I with different colour intensities were subjected to the Ames test using Salmonella typhimurium strains TA98, TA100, TA1535, TA 1537, and TA1538. Caramel colour I was neither mutagenic nor cytotoxic, either with or without activation by rat liver S-9 fraction, at concentrations up to 20 μl per plate (Richold & Jones, 1980a,b).

Acute toxicity
No information available.

Short-term studies
Rats

Caramel colour I was administered to groups of 20 weanling female Wistar rats at dietary levels of 0, 15, or 30% for 8 weeks followed by a 4-week recovery period. Diarrhoea was observed in the treated animals and food efficiency was decreased, but the growth rate was normal. Haematological indices, in particular leucocyte counts, were normal throughout the study. The relative caecal weights were increased after eight weeks, but returned to normal by the end of the 4-week recovery period. Discolouration of the mesenteric lymph nodes was observed in animals of both treatment groups after eight weeks, but the discolouration diminshed during the recovery period. No other gross or microscopic pathological changes were reported (Sinkeldam & van der Heyden, 1976).

Long-term studies
No information available.

Observations in man
No information available.

Comments
Caramel colour I is free of the heterocyclic compounds associated with convulsant activity or depressed lymphocyte counts which occur in carmels prepared using ammonia or ammonium salts, and displays a low order of short-term toxicity. None of the data suggest a need to revise earlier JECFA recommendations.

EVALUATION
Estimate of acceptable daily intake for man
ADI "not specified".

REFERENCES

Richold, M. & Jones, E. (1980a). Ames metabolic activation test to assess the potential mutagenic effect of ETA-38-2H. Unpublished report No. FDC 8/80359 from Huntingdon Research Centre, Huntingdon, England. Submitted to WHO by International Technical Caramel Association.

Richold, M. & Jones, E. (1980b). Ames metabolic activation test to assess the potential mutagenic effect of ETA-38-1W. Unpublished report No. FDC 8/80361 from Huntingdon Research Centre, Huntingdon, England. Submitted to WHO by International Technical Caramel Association.

Sinkeldam, E.J. & van der Heyden, C.A. (1976).   Short-term feeding
    test with three types of caramels in albino rats. Unpub-
    lished   report   No.   R4789   from   CIVO/TNO,   Zeist,   The
    Netherlands. Submitted to WHO by International Technical
    Caramel Association.

CARAMEL COLOUR II

EXPLANATION
The twenty-fourth meeting of the Committee (Annex 1, reference 53) drew attention to the fact that caramel colour II has no ADI
since it is not included in the ADI of caramel colour I. This material
has not previously been evaluated by the Committee.

BIOLOGICAL DATA
Biochemical aspects
No information available.

Toxicological studies
Special studies on mutagenicity
Caramel colour II at concentrations of 2.5-20 µl/plate was
neither mutagenic nor cytotoxic in the Ames test using Salmonella
typhimurium strains TA98, TA100, TA1535, TA1537, and TA1538, with or
without metabolic activation by rat liver S-9 fraction (Richold &
Jones, 1980).

In a similar test, caramel colour II was neither mutagenic
nor cytoxic at concentrations of 50-5000 µg/plate (Richold et al.,
1984).

Caramel colour II was tested for potential mutagenic
activity based on induction of DNA repair (unscheduled DNA synthesis)
in cultured human epithelial (HeLa 53) cells. Caramel colour II was
incorporated in the culture medium at concentrations of 25-51,200
µg/ml and the test was performed on 2 occasions both in the presence
and absence of rat liver S-9 mix. In both tests, in the absence of the

S-9 mix a small but statistically significant increase in the number of silver grains over nuclei was observed at a concentration of 25,600 µg/ml; no significant increases were seen at higher concentrations in this test nor at any concentration in the repeat test (Allen & Proudlock, 1984).

Caramel colour II was not clastogenic to cultured Chinese hamster ovarian cells at concentrations of 500, 2500, or 5000 µg/ml either in the presence or absence of rat liver S-9 mix (Allen et al., 1984).

Acute Toxicity
No information available.

Short-term studies
Rats

Five groups of 20 male and 20 female weanling Fischer F-344 rats were given caramel colour II in drinking water at dose levels of 0, 4, 8, 12, or 16 g/kg b.w./day for 90 days. Body weights and food and water intakes were recorded weekly. Haematological examinations, blood bichemical studies, and urinalysis were performed during the seventh week and at termination on all animals. Animals were fasted overnight prior to bleeding from the orbital sinus under anaesthesia; the animals were also fasted during the 24-hour urine collection period. At necropsy, organs were weighed and all animals were examined macroscopically. Histopathological examinations were conducted on all animals in the control and high-dose groups and on animals from the low- and mid-dose groups that died or were sacrificed in extremis.

All animals showed a weight loss and reduced food intake between weeks 7 and 8, presumably due to the fasting prior to bleeding and during urine collection. In addition, caramel colour II caused a dose-related reduction in body-weight gain and in food and fluid intake in both sexes. Males treated with 8 g caramel colour II per kg b.w., and females treated with 4 and 8 g/kg b.w., had significantly lower food consumption than controls for 4 of 13 weeks, 3 of 13 weeks, and 6 of 13 weeks, respectively; mean food consumption, of both males and females treated with 12 and 16 g/kg b.w., was significantly lower than food consumption of their respective controls. All treated groups had lower mean fluid intake than their respective controls, usually significantly lower. Males receiving 12 and 16 g/kg b.w. had significantly lower mean body weights than controls from weeks 7-13 and weeks 4-13, respectively. Females treated with 8 g/kg b.w. had significantly lower mean body weights than controls from week 11-13 while, beginning at week 7, females receiving 12 and 16 g/kg b.w. had lower mean body weights than controls.

Occasional statistically-significant changes were seen in some haematological and clinical chemical parameters, but the mean values were within the normal range for Fischer 344 rats and were not considered to be of toxicological significance. There was a dose-related reduction in urinary volume and pH, and an increase in urine specific gravity.

At necropsy, treatment-related increases were observed in kidney weights and in full and empty caecum weights, but no significant histopathological changes were observed in these or other tissues. Dose-related staining of the gastrointestinal tract and mesenteric lymph nodes was noted, and deposits of yellow pigment were seen histopathologically in the caecal submucosa and mesenteric lymph nodes of the top-dose groups (the only treated animals examined comprehensively). Reactive hyperplasia was not associated with the pigment deposition.

The investigators concluded that the reduced body weights, food and fluid intakes, and increased kidney weights were due to water imbalance reflecting poor palatability of the drinking solution rather than toxic effects of caramel colour II per se (MacKenzie, 1985).

Comments
The above data are insufficient to evaluate caramel colour II for an ADI, but the substance has a low sub-chronic toxicity and no frankly pathological effects were seen in the 90-day study.

EVALUATION
Estimate of acceptable daily intake for man
No ADI allocated.

REFERENCES

Allen, J.A., Brooker, P.C., Birt, D.M., & McCaffrey, K.J. (1984). Analysis of metaphase chromosomes obtained from CHO cells cultured in vitro and treated with caramel colour (II). Unpublished report No. ITC 3A/84965 from Huntingdon Research Centre, Huntingdon, England. Submitted to WHO by International Technical Caramel Association.

Allen, J.A., & Proudlock, R.J. (1984). Autoradiographic assessment of DNA repair in mammalian cells after exposure to caramel colour II. Unpublished report No. ITC 2A/84750/2 from Huntingdon Research Centre, Huntingdon, England. Submitted to WHO by International Technical Caramel Association.

MacKenzie, K.M. (1985). 90-day toxicity study of caramel color (caustic sulfite process) in rats. Vols. I & II. Unpublished report No. 6154-105 from Hazleton Laboratories America Inc., Madison, WI, USA. Submitted to WHO by International Technical Caramel Association.

Richold, M. & Jones, E. (1980). Ames metabolic activation test to assess the potential mutagenic effect of ETA-38-3N. Unpublished report No. FDC 8/80356 from Huntingdon Research Centre, Huntingdon, England. Submitted to WHO by International Technical Caramel Association.

Richold, M., Jones, E., & Fenner, L.A. (1984). Ames metabolic activation test to assess the potential mutagenic effect of caramel colour II. Unpublished report No. ITC 1A/84708 from Huntingdon Research Centre, Huntingdon, England. Submitted to WHO by International Technical Caramel Association.

CARAMEL COLOUR III

EXPLANATION

In earlier evaluations, the presence of 4-methylimidazole in caramel colour III was noted, particularly since this compound was likely to be the causal agent of convulsions in cattle and sheep fed ammonia-treated molasses. However, at its twenty-first meeting, the Committee (Annex 1, reference 44) no longer considered this a cause for concern since the introduction of chemical specifications limits the concentration of 4-methylimidazole in caramel colour III. The twenty-first meeting identified the principal toxic effect of ammoniated caramels as the depression of circulating lymphocytes and total leucocytes for which a no-effect level had not been determined; consequently, the temporary ADI for caramel colour III was revoked.

BIOLOGICAL DATA
Biochemical aspects
Absorption, distribution, and excretion

In groups of 2-4 rats, the absorption of the colour-giving components of caramel was determined by faecal extraction. Recoveries varied widely for the 10 or 20% caramel solutions examined despite pretreatment for 100 days before testing. About one-third of the colour-giving components appeared to be absorbed, but no conclusions could be drawn regarding the absorption of colourless components (Haldi & Wynn, 1951).

Toxicological studies
Special study on 4-methylimidazole (4-MeI)

4-MEI has been shown to be the most likely toxic component in ammoniated molasses, its being a convulsant to rabbits, mice, and chicks at oral doses of 360 mg/kg b.w. (Nishie et al., 1969).

Mice

Male albino mice (20-25 g) were used to determine the median convulsive dose (CD_{50}) and the median lethal dose (LD_{50}) of a few imidazoles. The results are given in the following table:

Convulsant and lethal effects of imidazoles

	$CD_{50} \pm SE$ in mg/kg b.w.		$LD_{50} \pm SE$ in mg/kg b.w.	
	i.p.	oral	i.p.	oral
4-methylimidazole	155 + 5	360 + 18	165 + 3	370 + 15
1-methylimidazole	380 + 8.2	1,400 + 79	380 + 8.2	1,400 + 79
2-methylimidazole	500 + 12	1,300 + 70	480 + 18	1,400 + 114
imidazole	560 + 34	1,800 + 45	610 + 7.4	1,880 + 45

All the imidazoles tested produced varying degrees of tremor, running, restlessness, dialorrhea, Straub tail, opisthotonus and tonic extensor seizure that ended in death (Nishie et al., 1970).

Chickens

The CD_{50} and the LD_{50} of 4-MEI by i.p. injection in 1-day-old chicks were 174±10 mg/kg b.w. and 210±15 mg/kg b.w., respectively. Orally, the CD_{50} was 580±30 mg/kg b.w. and the LD_{50} was 599 + 50 mg/kg b.w. Doses of 100 mg/kg b.w. i.p. caused tremors, peeping, and spreading of the wings. Doses over 150 mg/kg b.w. i.p. caused opisthotonus, prostration with clonic leg movements, and terminal tonic extensor seizure (Nishie et al., 1970).

Special studies on reduction of total lymphocyte counts

Following the request in the twenty-first report of the Committee for information on the factor(s) responsible for the haematological effects of caramel colour III, a series of studies have been performed to investigate the mechanism(s) underlying these effects and the agent(s) responsible.

Rats

The effects of vitamin E, folic acid, pyridoxine and choline on the capacity of caramel colour III to reduce total lymphocytes in the blood of rats fed caramel colour III were examined. Four groups of 10 male weanling rats were fed Spratt's diet containing 8% caramel colour III and supplemented with vitamin E (100 mg/kg), folic acid (10 mg/kg), pyridoxine hydrochloride (10 mg/kg), or choline chloride (1000 mg/kg). Groups on Spratt's diet alone or on Spratt's diet with 8% caramel colour III without any supplement were used as controls. After

12 days, there was a marked reduction in total white blood cells and lymphocytes in rats fed the diet containing 8% caramel colour III and neutrophil counts were increased. Dietary supplementation with vitamin E, folic acid, or choline did not noticeably affect lymphocyte counts. However, rats receiving the diet supplemented with pyridoxine had white blood cells and lymphocytes in numbers similar to those fed the basal diet without caramel colour III. The basal diet was found to contain 2.3 mg/kg pyridoxine.

In a further study using CIVO stock diet (pyridoxine concentration 3 mg/kg), rats receiving 8% caramel colour III in the diet showed a reduction in total white blood cell and lymphocyte numbers; these changes were ameliorated by the addition of 10 mg/kg pyridoxine in the diet. In this study, neutrophils were not affected by the caramel colour III treatment, but the plasma pyridoxal phosphate levels were reduced by treatment (Sinkeldam et al., 1984).

In order to quantify the relationship between dietary pyridoxine and reduction of lymphocyte counts by caramel colour III, and to study the effects of age, two 14-day studies were performed, one in weanling, and the second in mature, Wistar rats; in other respects the protocols were identical. Groups of 10 male animals were fed diets containing approximately 2.5, 6, 12, or 24 ppm of pyridoxine. At each dietary level groups were given caramel colour III[1] in the drinking water at levels of 0, 1, 4, or 8%. A clear inverse dose relationship between the severity of lymphocyte depression and the pyridoxine level of the diet was observed. In weanling rats fed a diet containing 2.5 ppm pyridoxine, statistically significant lymphocyte reduction occurred at all caramel colour III levels on days 6 and 13. Groups fed diets containing 6, 12, or 24 ppm pyridoxine did not show statistically significant reductions in lymphocyte counts on day 6. However, on day 13, groups receiving 4 or 8% caramel colour III and 6 ppm pyridoxine in the diet had a statistically significant reduction in lymphocytes. On

---

[1]   This sample of caramel colour III contained 107 mg/kg 2-acetyl-4(5)-tetrahydroxybutylimidazole (THI) on a colour equivalent basis, 204 mg THI/kg on an 'as is' basis, and 295 mg THI/kg on a solids basis.

day 13 at a dietary level of 12 ppm pyridoxine only the 8% caramel colour III group had a statistically significant reduction in lymphocytes. There were no significant reductions in lymphocyte counts in animals fed 24 ppm pyridoxine at any level of caramel colour III intake. In mature rats fed 4 and 8% caramel colour III a statistically significant reduction in lymphocytes occurred at dietary pyridoxine levels of 12 ppm and lower on day 6. On day 13 lymphocytes were significantly decreased at caramel colour III doses 1, 4, and 8% in animals fed a diet containing 2.5 ppm pyridoxine. There were no significant decreases in lymphocytes at these dose levels of caramel colour III in animals fed 6, 12, or 24 ppm pyridoxine (Sinkeldam, 1981; 1982a).

Special studies on 2-acetyl-4(5)-tetrahydroxybutylimidazole (THI)

An isolation procedure was developed and a fraction of caramel colour III that contained the lymphocyte-depressing activity was isolated. The single component of this fraction that was responsible for the activity was identified as THI (Kroplien et al., 1984).

Rats

THI was administered to groups of 10 male weanling Wistar rats at levels of 0, 2, 5, or 20 ppm in drinking water for 7 days; a fifth group received a 1% solution of caramel colour III in drinking water as a positive control. THI produced a marked depression of lymphocyte counts at all dose levels and a dose-dependent increase in neutrophils. The lymphocyte-depressing potency of 2 ppm THI was comparable to that of 1% caramel colour III in the drinking water, indicating a level of approximately 200 mg/kg THI in the caramel sample. This compares with the result of chemical analysis of a sample of this batch of caramel colour III, which indicated a value of 204 mg/kg THI on an 'as is' basis (Sinkeldam, 1982b; Kroplien, 1984).

Special studies on carcinogenicity (see also long-term studies)

Mice

Three groups of 50 male and 50 female $B_6C_3F_1$ mice were given drinking water containing 0, 1.25, or 5% caramel colour III for 96 weeks followed by 8 weeks of drinking water without caramel colour; the animals were fed CFR diet ad libitum. The caramel colour III used in this study contained less than 25 mg/kg THI.

The animals were observed daily for abnormalities and mice showing signs of ill-health were isolated to be returned to the group if their condition improved but otherwise killed and autopsied. Individual body weights were recorded weekly for the first 14 weeks, then every-other-week. Food and water consumption were recorded over a 2-day period before each weighing. During week 104, fresh urine samples were obtained from all survivors and analysed for pH, protein, glucose, bilirubin, ketones, occult blood, and urobilinogen. At termination the animals were killed by exsanguination under ether anaesthesia and haematological examinations were performed, which consisted of measurements of haemoglobin concentrations, haematocrit values, erythrocyte counts, leucocyte counts, platelet counts, and differential leucocyte counts. At autopsy, gross findings were recorded and the following organs weighed: brain, heart, liver, spleen, adrenals, and testes or ovaries. Samples of these organs and of the salivary gland, trachea, lungs, thymus, lymph nodes, stomach, small intestine, pancreas, urinary bladder, pituitary, thyroid, prostate, seminal vesicle, uterus, mammary gland, skeletal muscle, eye, Harderian glands, spinal cord, sciatic nerve, and any other tissues of abnormal appearance were examined histologically (haematoxylin- and eosin-stained). Histopathological examinations were also performed on mice that died and on those that were killed in moribund condition. During the in-life phase no consistent differences were noted between the test and control groups with respect to growth or water intake.

The cumulative mortality of males given 5% caramel colour III was higher than that of controls from week 100 to the end of the experiment, but there were no clear pathological differences in any organs and no treatment-related abnormalities in urinalyses.

The only statistically significant differences in haematological parameters between control and treated groups were elevations of the total leucocyte counts in males of both treatment groups, but the observed values were within the range encountered for the strain of $B_6C_3F_1$ mice used in the study. No treatment-related gross pathology was noted during or at the end of the experiment. Malignant lymphoma/leukaemia, hepatocellular carcinoma, and sub-cutaneous fibrosarcoma and/or malignant fibrous histocytoma were frequent, but no significant differences were found in their incidences between treated and control groups. Adenomas and adenocarcinomas of the lungs, hyperplastic nodules of the liver, and fibroma of the sub-cutis were frequent in males, but their incidences were similar in treated and untreated mice.

The authors concluded that, under the conditions used, caramel colour III was not carcinogenic for $B_6C_3F_1$ mice (Hagiwara et al., 1983).

Rats
Three groups of 50 male and 50 female F344 rats were given caramel colour III in drinking water at levels of 0, 1, or 4% for 104 weeks followed by drinking water without caramel for 9 weeks. The animals were fed ad libitum basal diet that contained 11-12 mg/kg pyridoxine, and the caramel colour III used contained less than 25 mg/kg THI. During the experimental period, all animals were observed daily, and clinical signs and mortality were recorded. Body weights were recorded weekly during the first 13 weeks of the study and then every 4 weeks. Moribund or dead animals and animals sacrificed at termination were autopsied and examined for the development of tumours in the following organs and tissues: brain, pituitary, thyroid (including parathyroid), thymus, lungs, trachea, heart, salivary glands, liver, spleen, kidneys, adrenals, tongue, oesophagus, stomach, duodenum, jejunum, ileum, caecum, colon, rectum, urinary bladder, lymph nodes, pancreas, gonads, accessory genital organs, mammary gland, skin, musculature, peripheral nerve, spinal cord, sternum, femur, eyes, ear duct, and nasal cavity. These tissues were also examined histologically (haematoxylin- and eosin-stained).

No treatment-related differences in growth or survival rates were noted. No dose-related effects were found either in the incidence or induction time of tumours in the various organs and tissues except in the pituitary gland of males of the top-dose group in which the incidence of tumours was significantly higher than that in controls. However, pituitary tumours are among the most common spontaneous tumours in F344 rats that occur with variable incidence; most of the tumours were microscopic and there were no significant differences in their induction times compared with controls. The authors concluded that the higher incidence of pituitary tumours was not related to caramel administration, but could be explained by the variability of spontaneous tumour incidence (Maekawa et al., 1983).

Special studies on mutagenicity

Caramel colour III (a blend of 3 commercial samples) was evaluated by the Ames Salmonella/microsome plate test and the Saccharomyces/microsome plate test. The Salmonella test organisms used were TA98, TA100, TA1535, TA1537, and TA1538; the Saccharomyces was strain D4, and the tests were conducted in the presence and absence of liver S-9 mix from Araclor-induced rats. The dose range employed was 1-50 mg/plate for TA100 and 1-20 mg/plate for all other test strains. Caramel colour III was non-mutagenic to the test organisms under these conditions (Jagannath & Brusick, 1978a).

In a similar study using the same Salmonella tester strains, caramel colour III was non-mutagenic in a range of concentrations up to 20 µl/plate with or without metabolic activation (Richold & Jones, 1980).

Caramel colour III was non-mutagenic in the Ames test against Salmonella strains TA98 and KTA100 at concentrations of 0-1 mg/plate, with or without metabolic activation. Acidic and basic organic extracts of the caramel were also non-mutagenic (Ashoor & Monte, 1983).

Thirteen commercial caramel colours (not identified) were examined for mutagenicity in the Ames test, using Salmonella strains

TA98 and TA100 with and without metabolic activation, and for DNA-damaging effects in <u>E. coli</u> (Wild/pol A⁻, Wild/rec A⁻); none of the samples tested was active under the test conditions (Kawana <u>et al</u>., 1980).

Five samples of commercial caramel colour (not identified) were tested for mutagenicity in the Ames test using <u>Salmonella</u> strains TA98 and TA100. All samples gave equivocal results with TA100 and 2 of the 5 samples were equivocal with TA98 (Kawachi <u>et al</u>., 1980).

Five samples of caramel colour were tested in the Ames <u>Salmonella</u> plate assay using strains TA98, TA100, and TA1537 without metabolic activation but with a 20-minute preincubation step. The same samples were used in chromosome aberration tests performed on a cultured Chinese hamster lung fibroblast cell line, both in the presence and absence of S-9 fraction. All samples of caramel colour were designated as positive both in the Ames assay and in the chromosome aberration test (aberrations in 20% of metaphase cells) (Ishidate & Yoshikawa, 1980).

The same series of caramel colour samples was tested by Jagannath and Brusick, and all were found to be non-mutagenic in the Ames test using <u>Salmonella</u> strains TA98, TA100, TA1535, TA1537, and TA1538, and <u>Saccharomyces</u> strain D4, in the presence or absence of S-9 mix. One of the samples in this study and in the Ishidate & Yoshikawa (1980) study was caramel colour III (Jagannath & Brusick, 1978b).

The mutagenicity of a series of samples taken at various stages in the manufacture of caramel colour III were assayed in the Ames test using <u>Salmonella</u> strains TA98, TA100, and TA1535, with and without metabolic activation. No mutagenicity was seen with any sample against all three tester strains in the presence of S-9 fraction but and increased number of revertants was found with strain TA100 in the absence of S-9 fraction. Mutagenic activity was associated with samples taken late in the process (Jensen <u>et al</u>., 1983).

A sample of caramel colour III was non-mutagenic in the Ames test against Salmonella strains TA98, TA100, TA1535, TA1537, and TA1538 both in the presence and absence of rat hepatic S-9 fraction at caramel coulour concentrations up to 5 mg/plate (Richold et al., 1984).

The same sample as that used by Richold et al. (1984b) was tested for potential mutagenic activity based on induction of unscheduled DNA synthesis in cultured human epithelial (HeLa 53) cells both in the presence and the absence of rat liver S-9 fraction. The test was performed on two occasions at caramel colour III concentrations of 25-51,200 μg/ml in the culture medium. In both tests, caramel colour III caused a significant increase in the number of silver grains found over cell nuclei at concentrations of 6,400 and 12,800 μg/ml in the absence of S-9 mix; significant increases were not observed at higher or lower concentrations in the absence of S-9 mix nor at any of the concentrations tested in the presence of S-9 mix. The authors concluded that, although reproducible effects on unscheduled DNA synthesis were demonstrated in the test, this effect may be moderated by metabolism (Allen & Proudlock, 1984).

Caramel colour III was not clastogenic to cultured Chinese hamster ovarian cells at concentrations of 500, 2500, or 5000 μg/ml either in the presence or absence of rat liver S-9 mix (Allen et al., 1984).

The clastogenicity of caramel colour III was evaluated in an in vitro cytogenetic assay using cultured Chinese hamster ovarian cells. A significant increase in chromosome aberrations was observed at concentrations of 3 mg/ml or higher in the absence of metabolic activation; similar findings were reported for sodium ascorbate at 2 mg/ml (Galloway & Brusick, 1981a).

These tests were repeated in the presence of rat liver S-9 fraction, and neither caramel colour III at a concentration up to 5 mg/ml nor sodium ascorbate at a concentration of 5 mg/ml was active (Galloway & Brusick, 1981b).

In an _in vivo_ mouse micronucleus test, caramel colour III was administered by gavage to 5 males and 5 females at doses of 0, 1.05, or 3.5 g/kg b.w. (2 doses 24 hours apart). Caramel colour III did not increase the incidence of micronuclei in polychromatic erythrocytes obtained from marrow and was considered by the authors not to exhibit clastogenic activity under the conditions of the test (Cimino & Brusick, 1981).

Special studies on teratogenicity

Teratogenicity studies were carried out on caramel colour III in mice, rats, and rabbits. The doses employed were 0, 16, 74.3, 345, and 1600 mg/kg b.w. in all three species.

Mice

Caramel colour III was administered by gavage to groups of 22 or 23 pregnant CD1 mice at the above doses beginning on day 6 and continuing through day 15 of gestation. On day 17, all dams were subjected to Caesarian section under anaesthesia and the numbers of implantation sites, resorption sites, and live and dead foetuses were recorded. The urogenital tract of each dam was examined for anatomical abnormalities and all foetuses were examined grossly for external abnormalities. One-third of the foetuses from each litter were examined for visceral abnormalities using the Wilson technique, and the remaining two-thirds were examined for skeletal defects after clearing and staining with Alizarin Red. Caramel colour III had no clearly discernible effects on nidation nor on maternal or foetal survival. The number of abnormalities of either soft or skeletal tissues of the test groups did not differ from those occurring spontaneously in controls (Morgareidge, 1974).

Rats

Caramel colour III was administered by gavage to groups of 21-24 pregnant Wistar rats at the above doses beginning on day 6 and continuing daily through day 15 of gestation. On day 20, all dams were subjected to Caesarian section under anaesthesia and the numbers of implantation sites, resorption sites, and live and dead foetuses were recorded. The body weights of live pups were also recorded. All

foetuses were examined grossly for external abnormalities. One-third
of the foetuses from each litter were examined for visceral abnormalities using the Wilson technique, and the remaining two-thirds were
examined for skeletal defects after clearing and staining with Alizarin
Red. Caramel colour III had no clearly discernible effect on nidation
nor on maternal or foetal survival. The number of abnormalities seen
in either soft or skeletal tissues of the test groups did not differ
from those occurring spontaneously in controls (Morgareidge, 1974).

Rabbits

Caramel colour III was administered by gavage to groups of
11 or 12 pregnant Dutch-belted female rabbits at the above doses
beginning on day 6 and continuing daily through day 18 of preganancy.
On day 29, all does were subjected to Caesarian section under anaesthesia and the numbers of corpora lutea, implantation sites, resorption
sites, and live and dead foetuses were recorded. All foetuses were
examined grossly for the presence of external congenital abnormalities. The live foetuses from each litter were then placed in an incubator for 24 hours for evaluation of neonatal survival. Surviving pups
were sacrificed and all pups examined for visceral abnormalities by
dissection. All foetuses were then cleared and stained with Alizarin
Red and examined for skeletal defects. Caramel colour III treatment
had no effect on nidation nor on maternal or foetal survival. The
number of abnormalities in either soft or skeletal tissues of the test
groups did not differ from those occurring spontaneously in controls
(Morgareidge, 1974).

Acute toxicity

Species	Route	LD_{50} (mg/kg b.w.)	Reference
Rat	oral	> 2.3 ml ≃ 1,900	Foote et al., 1958
Rat	oral	> 25 ml ≃ 17,500	Chacharonis, 1960
Rat	oral	> 30 ml ≃ 20,400	Chacharonis, 1963

No treatment-related effects were detected during the observation of animals for 14 days after administration of single doses of
12 different caramels manufactured with ammonia or sulphate ammonia
catalysts (Foote et al., 1958; Chacharonis, 1960; 1963). Single doses
of caramel colour III of up to 10 g/kg b.w. in mice and 15 g/kg b.w. in
rabbits did not cause convulsions or others signs of distress
(Sharratt, 1971).

Short-term studies
Mice

In a 4- to 6-week study, albino Swiss mice (10 males and 10
females/group) were fed caramel colour III containing 830 ppm 4-methy-
limidazole at concentrations of 0, 1, 2, 4, 8, or 16% in the diet. No
influence of caramel colour III on appearance, behaviour, or food
intake was observed. Growth was decreased, especially in the third and
fourth weeks, in the groups fed 16% caramel colour III. The faeces of
the animals fed the higher dose levels were soft, tarry in appearance,
poorly-formed, and sticky or pasty in consistency.

In the males fed 16% caramel colour III, an increase of
neutrophilic leucocytes and a decrease in lymphocytes were observed.
The mean relative weights of the caeca (full and empty) were increased
at the 4, 8, and 16% dietary levels. No other remarkable findings were
observed on gross examination; histopathological examinations were not
conducted. No information is available on the THI content of the
sample of caramel colour III used in this study, nor on the dietary
levels of pyridoxine (Procter, 1976).

In a pilot study to select the dose levels for a carcinogenicity study, $B_6C_3F_1$ mice were fed caramel colour III for 13
weeks. No details of this study were reported (Hagiwara et al., 1983).

Rats

Four groups of 10 male and 10 female rats received either 0
or 10% of 2 different samples of caramel colour III in their diet for
90 days. Weight gains showed slight reductions compared with controls,
but food consumption was normal in all groups. No abnormalities were
noted regarding haematology, urinalysis, gross pathology, or histo-

pathology. No information is available on the THI content of the samples of caramel colour III used in this study (Charcharonis, 1963).

Four groups of rats received 0, 4, 8, or 16% caramel colour III in their diet for 3 months. No convulsions or other behavioural abnormality or signs of neurological damage were seen. No macroscopic pathological abnormalities were found in the central nervous system (Sharratt, 1971).

Six groups of 15 rats (CFE strain) of each sex were fed diets containing 4, 8, or 16% caramel colour III produced by either an "open" or a "closed" pan process for 13 weeks; a group of 25 rats of each sex served as controls. Body-weight gain was decreased at all dietary levels of both caramel colours. Haemoglobin concentrations were reduced at the highest dietary levels at week 6, while at the lower levels this effect was wless clear. After 13 weeks, the males at all dose levels had significantly decreased haemoglobin concentrations, but this effect occurred in females only at the 8 and 16% levels. In some groups there was a less consistent decrease in the total number of red blood cells at 6 and 13 weeks. The total number of leucocytes was significantly decreased and a lymphocytopenia was present at all dose levels at 6 and 10 weeks; at 13 weeks these changes were observed only in males. At necropsy, decreased weights of the thymus and spleen were observed at the 8 and 16% dose levels. The caecal weights were increased at the 8 and 16% dose levels compared with controls. Increased relative liver and kidney weights suggested an effect on these organs. Changes in the weights of other organs were considered to be related to the differences of body weight between the groups. The volumes of urine excreted during a 6-hour period without water or in the 2-hour period after a water load were lower than the control values. The latter differences were accompanied by higher values for the specific gravity of the urine. The histopathological study did not reveal treatment-related changes. No details of the THI content of the caramel colours used in this study were available (Gaunt et al., 1975).

Seven groups of 20 female Wistar rats were fed diets containing 0, 15, or 30% of 3 types of caramel colour, one of which was a sample of caramel colour III, for 8 weeks followed by a 4-week recovery period. The diets containing caramel colour III caused dose-related decreases in body weights and food efficiency, and also caused diarrhoea at the 30% dietary level. Leucocyte counts at 4 weeks were significantly increased at the highest dose level whereas at 8, 10, and 12 weeks no differences were observed between test groups and controls. The relative weights of the caeca of animals receiving caramel colour III, both filled and empty, were increased at 4 and 8 weeks but reverted to normal after the 4-week recovery period.

Gross examination at autopsy after 8 weeks revealed a slight, dose-related brown discolouration of the mesenteric lymph nodes in a few animals of each test group. After recovery periods of 2 or 4 weeks the discolouration was less intensive, but still visible. Microscopically, the lymph nodes of the test rats showed accumulation of pigment-laden macrophages, which was not noticeably diminished after withdrawal of the caramel for 2 or 4 weeks. No details of the THI content of the caramel colour III used in this study were available (Sinkeldam & van der Heyden, 1976a).

In a 10-week feeding study, groups of 15 male and 15 female Sprague-Dawley rats were fed diets containing caramel colour III at concentrations of 0, 1.25, 2.5, 5.0, 10.0, or 15.0%. In the animals receiving caramel colour III at levels of 5% or higher, the faeces became soft within two weeks and the water content of the faeces was higher than the faeces of controls. Body-weight gains were generally reduced in animals fed caramel colour III, particularly during the last 2-4 weeks of the study.

In rats fed caramel colour III, there were no changes in haemoglobin levels or erythrocyte counts; a significant reduction in lymphocyte counts and a coincident increase in the number of segmented neutrophils was observed at all dose levels. No macroscopic evidence of abnormal pigmentation of the mesenteric lymph nodes was found. An increase in empty caecal weight was consistently evident in animals of both sexes given caramel colour III at the 5, 10, and 15% levels.

Histopathological examination did not reveal changes in the structure of the ileal or caecal mucosa nor in the reticuloendothelial components of the central or peripheral systems. No abnormal pigmentation of the lymph nodes was found. No details of the THI content of the caramel colour used in this study were available (Procter et al., 1976).

Groups of 10 male and 10 female Wistar rats were fed diets containing caramel colour III at concentrations of 0, 1.25, 2.5, 5.0, 10.0, or 15.0% for 10 weeks. Caramel colour III caused loose stools, particularly at the 5% dietary level and body weights were slightly decreased in both sexes at this level. Leucocyte (lymphocyte) counts were decreased in males and in females fed 15% caramel colour III; at lower dose levels this effect occurred only in females. The relative weight of the caecum (both full and empty) was increased by feeding caramel colour III. Minimal amounts of pigment were observed in mesenteric lymph nodes of several rats fed 1.25% and higher levels of caramel colour III. From this experiment it appeared that caramel colour III was more active than two other types of caramel tested concurrently in regard to growth depression, enlargement of the caecum, and decrease in leucocyte counts. No details of the THI content were available (Sinkeldam & van der Heyden, 1976b).

In a 10-week study, weanling Wistar rats were fed caramel colour III in the diet at levels of 0, 0.5, 1.0, 2.0, 4.0, or 16%. There were 15 males and 15 females in each group except for the control group, which had 60 animals of each sex, and the highest dose group, which had 10 animals of each sex. Food intake and growth rates were recorded, and haematological examinations and histopathological studies were carried out. In particular, the lymph nodes, thymus, spleen, and caecum were examined for distribution of pigment.

Another 2 groups of 10 rats were given basal diet or diet containing 16% caramel colour III for 10 weeks followed by a 28-day period during which the basal diet was fed (recovery experiment). Five rats of each sex were killed after 7 days and the remainder after 28 days of the recovery period.

Caramel colour III depressed body-weight gain at dietary levels greater than 1%. Total leucocyte counts were decreased in males in the groups receiving 2, 4, and 16% caramel colour III and in females receiving 4 and 16% caramel colour III. However, the lymphocyte/

neutrophil ratio was significantly decreased in both sexes at all dose levels. Relative liver weights were increased at dietary levels of 2% and higher levels of caramel colour III, and reduced spleen weights were observed at the highest (16%) dose level. Increased relative kidney weights were observed in both sexes at 2% and higher dietary levels of caramel colour III. Increased caecal weights were observed at the 16% dietary level; microscopically, pigment was observed in the mesenteric lymph nodes of male and female rats at this dose level.

During the recovery phase, white cell counts, cell ratios, and total numbers of lymphocytes rapidly returned to normal; recovery was complete in both sexes by 7 days. The caecal weights had returned to normal by 7 days and the relative liver and spleen weights returned to normal during the recovery phase, while kidney weights partially recovered. No details of the THI content of the caramel nor of the pyridoxine content of diet were available (BIBRA, 1977).

Test groups of 15 male and 15 female Sprague-Dawley rats were fed diets containing 10 or 15% caramel colour III for 4 weeks; a control group of 20 rats of each sex received basal diet. During the course of the study, the rats fed diets containing caramel colour III had soft dark-coloured faeces, particularly at the highest dietary level. There were no consistent differences in body-weight gain or food consumption and no mortality occurred. Haematological studies were conducted prior to feeding caramel colour III and after 2 and 4 weeks. In males, total white cell and lymphocyte counts were unaffected by treatment at either sampling interval or dose level. In females, total white cell and lymphocyte numbers were significantly depressed after 4 weeks but not at 2 weeks at both dose levels. However, significant differences in differential white cell counts were observed in both sexes. In males, the differential lymphocytes (%) were significantly depressed at both dose levels and treatment intervals, and there was a concomitant increase in segmented neutrophils. In females, the differential counts of lymphocytes and segmented neutrophils were similarly affected after 4 weeks.

At necropsy, increased caecal weights were observed in both sexes and a statistically significant increase in relative weights of the thymus in male rats fed both doses of caramel colour III was noted. Histopathological studies were not conducted. Details of the THI content of the caramel sample used were not available (Procter, 1977).

In a 13-week toxicity study, caramel colour III was administered in drinking water at concentrations of 0, 4, 6, 8, or 10% to groups of 10 male and 10 female weanling Wistar rats. The caramel colour III sample used was analysed and found to contain 78 mg/kg THI on an 'as is' basis, 105 mg/kg THI on a solids basis. The diet fed to the rats contained approximately 13 mg/kg pyridoxine. The general condition and behaviour of the animals was checked at frequent intervals, individual body weights were measured weekly, food consumption was measured (on a cage basis) over weekly periods during the whole experimental period, and fluid consumption was measured daily (on a cage basis). Samples of blood for haematology were collected from the tip of the tail of all rats initially and at days 29/30, 57/58, and 84/85. Urinalysis was performed on individual urine samples from all animals during the last 16 hours of a 24-hour period of deprivation of foood and water at day 87. Tail tip blood collected at day 87 was examined for glucose and urea nitrogen. Blood was obtained from the aorta under ether anaethesia at termination (day 91 for males, 92 for females), and the following assays performed on the plasma: alkaline phosphatase, GPT, GOT, lactate dehydrogenase, total protein, and albumin. Pyridoxal phosphate was measured in plasma from EDTA-treated blood samples. At autopsy the following organs were weighed: thyroid, adrenals, testes/ovaries, kidneys, thymus, brain, spleen, heart, liver, and caecum (full and empty). Histological examinations were carried out on all animals of the control and top-dose groups and, in addition to the weighed organs, the following organs/tissues were examined: aorta, axillary and mesenteric lymph nodes, cervix, colon, oesphagus, stomach, duodenum, ilium, jejunum, lungs, epididymides, pituitary, prostate, skeletal muscle, skin, sternum, pancreas, trachea, urinary bladder, uterus, and all gross lesions.

There was a dose-related decrease in fluid consumption, decreased urinary output of more concentrated urine, decreased food consumption, and decreased body-weight gain in both sexes. These changes were related to the palatability of the drinking fluid. No outstanding differences were observed in red blood cell analyses between test groups and contols. Total lymphocyte counts were relatively low in all test grops of both sexes. The differences from control values were statistically significant in all male treatment groups after 29/30 days, in the female 8% group after 29/30 days, and in the male 4 and 8% groups after 57/58 days. However, these differences did not show a clear dose-dependent relationship and at 13 weeks there were no significant differences among any of the treatment groups compared to controls.

At necropsy, the relative weights of the kidneys and caecum were increased in several groups receiving caramel colour III. The increase in kidney weight was dose-dependent, but no treatment-related histopathological changes were seen in any of the organs examined and the enlargement of the kidneys was attributed to the decreased fluid consumption by the treated rats. No effects on the relative weights of spleen or thymus were observed and these tissues were histologically normal (Sinkeldam et al., 1980b).

A 13-week toxicity study was conducted with caramel colour III given in the drinking water at concentrations of 0, 5, 10, 15, or 20% to groups of 10 male and 10 female weanling Wistar rats. The caramel colour III sample used in this study contained 0-3 mg/kg THI on an 'as is' basis. The diet fed to the rats contained 13.5 mg/kg pyridoxine. The protocol used was similar to that described in the previous study except that blood samples were collected at days 30/31, 57/58, and 80/81, and urinalysis was performed at day 85.

Water intake (fluid intake corrected for caramel colour solids content) showed a dose-related decrease in both sexes. Food intake was also generally lower in all treated groups. Body-weight gains of males were decreased in a dose-related manner, while body-weight gains of females were comparable to controls. Urine volumes were decreased in rats treated with caramel colour III and the urines were more concentrated in treated groups than in contols. These changes were attributed to the decreased water intake. The urine was darker-coloured at the 15 and 20% dose levels; however, urine composition was essentially normal. Lymphocyte counts were in general relatively low in all test groups in both sexes but the differences only attained statistical significance in males of the top-dose group after 30 days and males of all treatment groups after 57 days. No significant differences in lymphocyte counts were observed in males of any group after 80 days, nor in females at any dose level at any time interval. Mean neutrophil counts were significantly increased in females receiving 20% caramel colour III for 81 days.

Although the relative weights of the caecum, liver, brain, kidneys, and testes were increased in several test grops, there were no pathological changes in any of these organs. Treatment-related microscopic changes consisting of increased numbers of macrophages containing a yellow-brown PAS-positive pigment were found in the mesenteric lymph nodes. These were the only treatment-related changes noted (Sinkeldam et al., 1980a).

Groups of 10 male and 10 female weanling F344 rats were given caramel colour III in the drinking water at concentrations of 0, 0.5, 1.0, 2.0, 4.0, or 8.0% for 4 weeks. The sample of caramel colour III used was found to contain 70 mg/kg THI on an 'as is' basis and the NIH 07 Open Formula Mouse and Rat diet used in this study contained 17 mg/kg pyridoxine.

No differences in body-weight gain were noted for any of the test groups, although food consumption of the males was significantly lower than controls throughout the study. Transient decreases in lymphocyte counts were noted among the treatment groups at the midpoint of the study, but no significant differences between control and treated groups were noted at 1 month. Other haematological and clinical chemistry parameters were normal for all groups. At necropsy no differences in organ weights were noted between treated and control groups and there were no gross or microscopic pathological changes related to treatment (Heidt and Rao, 1981).

Groups of 10 male and 10 female F344 rats were given caramel colour III in the drinking water at concentrations of 0, 1.25, 2.5, 5.0, 10.0, or 20.0% for 13 weeks in order to select doses for long-term carcinogenicity/chronic toxicity studies. Rats were given 20 ml/day of these solutions and basal diet (CRF-1, Charles River Japan, Inc.) was available ad libitum. The caramel colour III used contained 78 mg/kg

THI on an 'as is' basis and the diet contained 11-12 mg/kg pyridoxine. During the experimental period, all animals were observed daily and clinical signs were recorded. Body weights were measured every other week and haematological examinations were performed every 4 weeks in the control and 20% groups. At the end of the study, all survivors were sacrificed for gross and microscopic examination.

Weight gains were less in all experimental groups than in the control group from the first experimental week, and in week 13 the weight gains of the 1.25, 2.5, 5.0, 10.0, and 20.0% groups were 89, 94, 84, 76, and 76%, respectively, of that of controls for males and 96, 98, 80, 84, and 92%, respectively, for females. Except in the male 2.5% group and the female 1.25, 2.5, and 20.0% groups, the differences in weight gain from the controls were significant. No haematological changes were observed either during or at the end of the experimental period. At necropsy, no pronounced macroscopic changes were observed in any animals, although a few rats in the experimental groups were very emaciated. No histological changes related to caramel colour III administration were found in any experimental groups.

From these results, the authors concluded that 1 and 4% caramel colour III in drinking water were the appropriate dose levels for a carcinogenicity study (Maekawa et al., 1983).

In a 90-day toxicity study, 2 samples of caramel colour III were used, one containing approximately 15 mg/kg THI on a solids basis (batch A) and the other containing 295 mg/kg THI on a solids basis (batch B). Groups of 20 male and 20 female weanling F344 rats were given caramel colour III in drinking water at dose levels of 0, 10, 15, or 20 g/kg b.w. of batch A and 20 g/kg b.w. of batch B. The NIH 07 Open Formula Mouse and Rat diet used in this study contained more than 10 mg/kg pyridoxine. During the study, body weights and food intake were recorded weekly. Fluid consumption was recorded 3 times per week and concentrations of the test material were adjusted on the basis of body weight and fluid intake to give the required dose. Haematological analyses were performed at 2 and 6 weeks and at termination, and clinical chemistry analyses were carried out at 6 weeks and termination. At necropsy, all animals were examined macroscopically and selected organ

weights were determined. Histopathological examinations were conducted
on the 20 g/kg test groups and on the control animals.

Although the findings were not always consistent, the
animals treated at the higher dose levels (15 and 20 g/kg) of batch A
and those teated with batch B generally had decreased body weights.
All treated groups had significantly decreased food and fluid intake.
If fluid intake values are corrected for solid contents of the caramel
colours, the values for water intake were markedly below those of the
controls. Because of the rather constant fluid intake of the treated
groups, it was necessary to increase progressively the caramel colour
concentration to maintain a constant intake of the caramel colour on a
body-weight basis. It is likely that the effects on body-weight gain
and food consumption were due to the reduced water intake of the rats,
reflecting the poor palatability of the drinking solution rather than
toxic effects of the caramel colour per se.

The haematology studies revealed decreased lymphocyte counts
in the male and female rats fed batch B at 2 weeks and in the male rats
fed batch B at 6 weeks. All lymphocyte values in these groups were
normal at the termination of the study. No decreases in lymphocyte
counts occurred in male or female groups fed batch A at any of the dose
levels. There were no consistent changes in clinical chemistry values,
with the exception of slightly increased values in blood urea nitrogen
in the rats treated with batch B. Urinalysis revealed decreased
urinary volume and increased specific gravity in rats of both sexes
treated with either batch of caramel colour III at 6 weeks. At
termination, these differences were only significant in male rats in
the top-dose groups (both batches). Treatment-related increases were
observed in the absolute and relative weights of the caecum (full and
empty) in animals of both sexes at all dose levels. Dose-related
increases occurred in absolute and relative kidney weights and were
considered to reflect compensatory hypertrophy as a consequence of
reduced water intake; there were no histophathological changes in the
kidneys of any of the test groups. A dose-related decrease in the
absolute weight of the thymus was observed, which reached statistical
significance in the top-dose group males with both samples of caramel
colour III and top-dose group females with batch B; this decrease was
not evident when the thymus weight was expressed relative to body

weight. Other differences in absolute and relative organ weights appeared to be a consequence of the dose-related reduction in body weight.

The only treatment-related microscopic changes noted were minimal to moderate accumulation of pigment in the tissues of the intestinal tract and mesenteric lymph nodes without pathologic alteration.

Although statistically-significant changes were identified in some clinical and anatomical parmeters, the authors did not consider them to be toxicologically important. On this basis, the no-adverse-effect level for ammonia caramel III was considered to be 20 g/kg b.w. (MacKenzie, 1985).

A paired-feeding study was conducted to determine whether poor palatability is the mechanism underlying the decreased body-weight gains frequently noted in toxicity studies of caramel colour III. A group of 10 male Wistar rats were fed 12% caramel colour III in the drinking water, and similar groups were permitted a limited intake of food or water equivalent to that consumed by the caramel colour III group. Body-weight gain and food and fluid intake were decreased in the group fed caramel colour III. A similar decrease in body-weight gain was noted in the groups of rats restricted to the equivalent intake of either food or water. The author concluded that the growth depression observed in rats when caramel colour III is fed in the drinking water is the result of decreased fluid and food intake. Poor palatability of drinking fluid containing caramel colour III was the probable cause of these changes (Sinkeldam, 1979).

Long-term studies
Rats
Four groups of 48 male and 48 female Wistar rats were given diets containing 0, 1, 3, or 6% caramel colour III (ammonia catalysed "half open-half closed pan" caramel; no indication about the presence of 4-methylimidazole was given) for 2 years. Food and water intake, growth, mortality, and organ weights were measured; haematological examinations, urinalyses, kidney function tests, and histopathological examinations also were carried out.

A decrease in growth, which was significant in the males, was observed at all dose levels. This effect was accompanied by a reduction in the cumulative food intake. A significant reduction in white cell number (in the 6% group) was associated in the early part of the study with a lymphocytopenia, which was present until week 80 in the male rats fed a diet containing 3 or 6% caramel. In the female animals the lymphocytopenia was present until week 52 in these 2 groups (the 1% group was not tested).

Spleen weights were reduced in a dose-related manner. The relative weights of the (full) caeca were clearly increased at all dose levels. No changes were found in the pancreata of the control animals, while in the test groups (not dose-related) a total of 10 hyperplastic changes were found. However, the number of tumours of the pancreas showed no relation to the administration of caramel. There was no evidence of a carcinogenic effect.

The authors concluded that a no-observed-effect level could not be established for the carmel used in this particular study (Evans et al., 1976).

Observations in man

In a number of animal studies with caramel colour III, a decrease in the total number of leucocytes associated with a decrease in the number of lynphocytes was noted. A pilot study in humans was carried out in which 1.5 g of caramel colour III (prepared by a closed-pan process) was ingested daily by 9 volunteers for 21 days. Total circulating leucocytes, lymphocytes, and erythrocytes, together with haemoglobin concentrations, were measured prior to and during the treatment. No changes were found that could be attributed to treatment with caramel colour III. In this experiment 6 of the subjects showed no differences from normal in stool frequency or condition. Three volunteers occasionally had soft stools; no control groups were used (BIBRA, 1976).

Comments

The temporary ADI for caramel colour III was revoked at the twenty-first meeting of the Committee due to its effects on circulating total leucocytes and lymphocytes. Since that time, a component of caramel colour III, 2-acetyl-4(5)-tetrahydroxybutylimidazole (THI), has been identified and shown to cause a depression of lymphocyte counts; the lymphocyte depression caused by caramel colour III has been shown to be due largely, if not solely, to this minor component. Comparison of the lymphocyte-depressing activity of pure THI with a batch of caramel colour III containing a known level of THI indicated that other components of this batch had an insignificant activity. The lymphocyte depression was largely ameliorated by dietary pyridoxine. In studies on samples of caramel containing 10 mg/kg THI, no effects on lymphocyte counts were observed with adequate dietary levels of pyridoxine.

Long-term studies in rats and mice indicate that caramel colour III is not carcinogenic at dose levels of up to 4% in drinking water, which was also the no-effect level in these long-term studies.

EVALUATION
Level causing no toxicological effect

As it was not possible to include caramel colour III at higher levels than 4% in drinking water in long-term studies, and as the effect of most concern, i.e. lymphopenia, could best be evaluated from short-term studies, the Committee based its evaluation on the no-effect level of 20 g/kg b.w./day in a 90-day study in rats using caramel colour III which contained approximately 15 ppm THI on a solids basis (10 ppm on an 'as is' basis).

Estimate of acceptable daily intake for man
0-200 mg/kg b.w. (0-150 mg/kg b.w. on a solids basis).

Further work or information
Desired

Analytical data to confirm that the sample on which the evaluation is based is representative of current commercial samples.

Studies to confirm that THI is the sole component of caramel colour III that has lymphocyte-depressing activity and to establish a no-effect level for THI.

REFERENCES

Allen, J.A., Brooker, P.C., Birt, D.M., & McCaffrey, K.J. (1984). Analysis of metaphase chromosomes obtained from CHO cells cultured in vitro and treated with caramel colour III. Unpublished report No. ITC 3B/84966 from Huntingdon Research Centre, Huntingdon, England. Submitted to WHO by International Technical Caramel Association.

Allen, J.A. & Proudlock, R.J. (1984). Autoradiographic assessment of DNA repair in mammalian cells after exposure to caramel colour III. Unpublished report from Huntingdon Research Centre, Huntingdon, England. Submitted to WHO by International Technical Caramel Association.

Ashoor, S.H. & Monte, W.C. (1983). Mutagenicity of commercial caramels. Cancer Letters 18, 187–190.

BIBRA (1976). A study of the haematological effects of caramel in human volunteers. Unpublished report No. 1/172/76 from the British Industrial Biological Research Association, Carshalton, Surrey, England.

Chacharonis, P. (1960). Acute and chronic toxicity studies on caramel colours A and B. Unpublished report No. S.A. 54219 from Scientific Associates Inc., St. Louis, MO, USA.

Chacharonis, P. (1963). Acute oral toxicity study in rats on caramel colorings 25A-1, 30B-0, and 30F-1. Unpublished report No. S.A. 79105 from Scientific Associates Inc., St. Louis, MO, USA.

Cimino, M.C. & Brusick, D.J. (1981). Mutagenicity evaluation of ETA-48-IH caramel color in the mouse micronucleus test. Unpublished report No. 22129 from Litton Bionetics Inc., Kensington, MD, USA. Submitted to WHO by International Technical Caramel Association.

Evans, J.G., Butterworth, K.R., Gaunt, I.F., & Grasso, P. (1976). Long-term toxicity study in the rat of a caramel produced by the "half-open-half closed pan" ammonia process. Unpublished report No. 6/1976 from the British Industrial Biological Research Association, Carshalton, Surrey, England.

Foote, W.L., Robinson, R.F., & Davidson, R.S. (1958). Toxicity of caramel color products. Unpublished report of Battelle Memorial Institute, Columbus, OH, USA.

Galloway, S.M. & Brusick, D.J. (1981a). Mutagenicity evaluation of ETA-48-IH in an in vitro cytogenetic assay measuring chromosome aberration frequencies in Chinese hamster ovary (CHO) cells. Unpublished report No. 20990 from Litton Bionetics Inc., Kensington, MD, USA. Submitted to WHO by International Technical Caramel Association.

Galloway, S.M. & Brusick, D.J. (1981b). Mutagenicity evaluation of ETA-48-IH in an in vitro cytogenetic assay measuring chromosome aberration frequencies in Chinese hamster ovary (CHO) cells. Part II. Unpublished report No. 20990 from Litton Bionetics Inc., Kensington, MD, USA. Submitted to WHO by International Technical Caramel Association.

Gaunt, I.F., Lloyd, A.G., Grasso, P., Gangolli, S.P., & Butterworth, K.R. (1975). Toxicological investigations of caramels. I. A short-term study in the rat with two caramels produced by variations of the ammonia process. Unpublished report No. 14 from the British Industrial Biological Research Association, Carshalton, Surrey, England.

Hagiwara, A., Shibata, M., Kurata, Y., Seki, K., Furushima, S., & Ito, N. (1983). Long-term toxicity and carcinogenicty test of ammonia-process caramel colouring given to $B_6C_3F_1$ mice in the drinking water. Fd. Chem. Toxicol. 21, 701-706.

Haldi, J., & Wynn, W. (1951). A study to determine whether or not caramel has any harmful physiological effect. I. Unpublished report from Emory University, Atlanta, GA, USA.

Heidt, M. & Rao, G.N. (1981). Subchronic toxicity study of ammonia caramel color type AC2 in rats. Unpublished report No. 80036 from Raltech Scientific Services, Inc., Madison, WI, USA. Submitted to WHO by International Technical Caramel Association.

Ishidate, M. Jr. & Yoshikawa, K. (1980). Chromosome aberration tests with Chinese hamster cells in vitro with and without metabolic activation - a comparative study on mutagens and carcinogens. Arch. Toxicol. Suppl. 4, 41-44.

Jagannath, D.R. & Brusick, D. (1978a). Mutagenicity evaluation of ETA 4-10, ETA 4-11, ETA 4-15 in the Ames Salmonella/microsome plate test. Unpublished report No. 20838 from Litton Bionetics Inc., Kensington, MD, USA. Submitted to WHO by International Technical Caramel Association.

Jagannath, D.R. & Brusick, D. (1978b). Mutagenicity evaluation of ETA 4-8, ETA 4-9, ETA 4-12, ETA 4-13, ETA 4-14 in the Ames Salmonella/microsome plate test. Unpublished report No. 20838 from Litton Bionetics Inc., Kensington, MD, USA. Submitted to WHO by International Technical Caramel Association.

Jensen, N.J., Willumsen, D., & Knudsen, I. (1983). Mutagenic activity at different stages of an industrial ammonia caramel process detected in Salmonella typhimurium TA 100 following pre-incubation. Fd. Chem. Toxicol. 21, 527-530.

Kawachi, T., Yahagi, T., Kada, T., Tazima, Y., Ishidate, M., Sasaki, M., & Sugiyama, T. (1980). Cooperative programme on short-term assays for carcinogenicity in Japan. In Molecular and Cellular Aspects of Carcinogen Screening Tests, IARC Scientific Publication No. 27, 323-330. International Agency for Research on Cancer, Lyon, France.

Kawana, K., Akema, R., Nakaoka, T., Ikeda, H., Takimoto, T., & Kawauchi, S. (1980). Studies on mutagenicities of natural food additives. II. Mutagenicities and antibacterial activities of caramels. Eisei Kagaku 26, 259-263.

Kroplien, U. (1984). Letter dated 17 December 1984 to International Technical Caramel Association, submitted to WHO.

Kroplien, U., Rosdorfer, J., van der Greef, J., Long., R.C. Jr., & Goldstein, J.H. (1985). 2-Acetyl-4(5)-(tetrahydroxybutyl)-imidazole: Detection in commercial caramel colour III and preparation by a model browning reaction. J. Org. Chem. 50, 1131-1133.

MacKenzie, K.M. (1985). 90-day toxicity study of caramel color (ammonia process) in rats. Vols. I & II. Unpublished report No. 6154-107 from Hazleton Laboratories America, Inc., Madison, WI, USA. Submitted to WHO by International Technical Caramel Association.

Maekawa, A., Ogiu, T., Matsuoka, C., Onodera, H., Furuta, K., Tanigawa, H., Hayashi, Y., & Odashima, S. (1983). Carcinogenicity study of ammonia-process caramel in F344 rats. Fd. Chem. Toxicol. 21, 237-244.

Morgareidge, K. (1974). Teratologic evaluation of FDA 71-82 (caramel, bakers and confectioners) in mice, rats and rabbits. Unpublished report No. PB 234-870 from Food and Drug Research Laboratories Inc., Waverly, NY; USA.

Nishie, K., Waiss, A.C., & Keyl, A.C. (1969). Toxicity of methylimidazoles. Toxicol. Appl. Pharmacol. 14, 301-307.

Nishie, K., Waiss, A.C., & Keyl, A.C . (1970). Pharmacology of alkyl and hydroxyalklpyrazines. Toxicol. Appl. Pharmacol. 17, 1.

Procter, B.G. (1976). A preliminary evaluation of the potential toxicological effects of ammonia caramel in mice. Unpublished report No. 5705 from Bio-Research Laboratories Ltd., Pointe Claire, Quebec, Canada.

Procter, B.G. (1977). A preliminary evaluation of the potential toxicological effects of ammonia caramel (U.S. origin) in rats. Unpublished report No. 5617 from Bio-Research Laboratories Ltd., Pointe Claire, Quebec, Canada. Submitted to WHO by International Technical Caramel Association.

Procter, B., Berry, G., & Chappel, C.I. (1976). A toxicological evaluation of various caramels fed to albino rats. Unpublished report No. 4244 from Bio-Research Laboratories Ltd., Pointe Claire, Quebec, Canada.

Richold, M. & Jones, E. (1980). Ames metabolic activation test to assess the potential mutagenic effect of ETA-48-1H. Unpublished report No. FDC 8/80358 from Huntingdon Research Centre, Huntingdon, England. Submitted to WHO by International Technical Caramel Association.

Richold, M., Jones, E., & Fenner, L.A. (1984). Ames metabolic activation test to assess the potential mutagenic effect of caramel colour III. Unpublished report No. ITC 1B/84709 from Huntingdon Research Centre, Huntingdon, England. Submitted to WHO by International Technical Caramel Association.

Sharratt, M. (1971). Unpublished report.

Sinkeldam, E.J. (1979). Paired feeding test in rats with caramel (AC_3) in drinking water. Unpublished report No. R 6157 from Centraal Instituut voor Voedingsonderzoek (CIVO/TNO), Zeist, The Netherlands. Submitted to WHO by International Technical Caramel Association.

Sinkeldam, E.J. (1981). Quantitative relationship between dietary pyridoxine content and lymphocyte counts in rats fed ammonia caramel. Unpublished report No. V81.390/211704 from Centraal Instituut voor Voedingsonderzoek (CIVO/TNO), Zeist, The Netherlands. Submitted to WHO by International Technical Caramel Association.

Sinkeldam, E.J. (1982a). Quantitive relationship between dietary pyridoxine content and lymphocyte counts in mature rats fed ammonia caramel. Unpublished report No. V82.102/220102 from Centraal Instituut voor Voedingsonderzoek (CIVO/TNO), Zeist, The Netherlands. Submitted to WHO by International Technical Caramel Association.

Sinkeldam, E.J. (1982b). Short-term (7 day) bioassay in rats with three dose levels of 2-acetyl-4(5)-tetrahydroxybutyl-imidazole (13-A-82). Unpublished report No. V82.291/221153 from Centraal Instituut voor Voedingsonderzoek (CIVO/TNO), Zeist, The Netherlands. Submitted to WHO by International Technical Caramel Association.

Sinkeldam, E.J., Bruyntjes, J.P., & Kuper, C.F. (1980a). Sub-chronic (13-week) oral toxicity study with a modified ammonia caramel (ETA-26-1) in rats. Unpublished report No. R 6483 from Centraal Instituut voor Voedingsonderzoek (CIVO/TNO), Zeist, The Netherlands. Submitted to WHO by International Technical Caramel Association.

Sinkeldam, E.J., Roverts, W.G., & Kuper, C.F. (1980b). Sub-chronic (13-week) oral toxicity study with ammonia caramel (AC_2) in rats. Unpublished report No. R 6121 from Centraal Instituut voor Voedingsonderzoek (CIVO/TNO), Zeist, The Netherlands. Submitted to WHO by International Technical Caramel Association.

Sinkeldam, E.J., de Groot, A.P., van den Berg, H., & Chappel, C.I. (1984). The effect of vitamin B_6 on the number of lymphocytes in blood of rats fed caramel colour (ammonia process). Unpublished report submitted to WHO by International Technical Caramel Association.

Sinkeldam, E.J. & van der Heyden, C.A. (1976a). Short-term feeding test with three types of caramels in albino rats. Unpublished report No. R 4789 from Centraal Instituut voor Voedingsonderzoek (CIVO/TNO), Zeist, The Netherlands.

Sinkeldam, E.J. & van der Heyden, C.A. (1976b). Short-term (10 week) feeding study in rats with three different ammonia caramels. Unpublished report No. R 5120 from Centraal Instituut voor Voedingsonderzoek (CIVO/TNO), Zeist, The Netherlands.

CARAMEL COLOUR IV

EXPLANATION

The report of the twenty-fourth meeting of the Committee (Annex 1, reference 53) drew attention to the need for adequate specifications for caramel colour IV and for a long-term study of carcinogenicity. The temporary ADI of 0-100 mg/kg b.w. was extended pending the results of long-term toxicity studies.

BIOLOGICAL DATA
Biochemical Aspects
Absorption, distribution, and excretion

Pigmentation of tissues of the lymphoreticular system, particularly of the mesenteric lymph nodes, has been a frequent observation in rats fed high levels of caramel colour. A study was undertaken to determine the distribution of the colour component of caramel colour in rats and to determine whether accumulation of caramel colour was the cause of pigmentation in the mesenteric lymph nodes.

Experimental batches of unlabelled and ^{14}C-labelled caramel colour IV were prepared and a concentrate of the higher molecular-weight colour components was made by ultrafiltration. The absorption, tissue distribution, and excretion of this fraction was studied in male F344 rats after a single oral gavage of 2.5 g/kg b.w. ^{14}C-Labelled material was administered to naive animals and to animals which had received 2.5 g/kg b.w. unlabelled material in drinking water daily for 13 days prior to administration of ^{14}C-labelled material.

Differences between the results in naive and pretreated animals were small. Most of the colour components were not absorbed, but instead were excreted in the faeces, mainly within 48 hours. By 96 hours after dosing, 99.7-102.4% of the administered dose was excreted in the faeces, with only 1-2% in urine and insignificant amounts (< 0.1%) as ^{14}CO$_2$ in expired air.

Groups of 4 rats were killed at intervals of 4, 8, 12, 24, and 96 hours after receiving the single oral dose of ^{14}C-labelled material, and radioactivity was measured in the following tissues/organs: blood, brain, heart, lungs, liver, kidneys, spleen, thymus, mesenteric and cervical lymph nodes, gastrointestinal tract (contents and tissues), and carcass. Most of the radioactivity was located in the gastrointestinal tract, and only low levels of radioactivity were found in blood and tissues. The specific activity in the thymus, mesenteric lymph nodes, spleen, kidneys, and liver exceeded that in blood, but with the exception of the mesenteric lymph nodes, the radioactivity was cleared rapidly over the 96-hour study period. With the exception of the gastrointestinal tract, the highest tissue levels attained, in the liver and kidneys, never exceeded 0.02% and 0.01% of the administered dose.

The results demonstrate that, after administration of large doses of the coloured components of caramel colour IV, only a small fraction was absorbed, distributed in lymphoreticular tissue, and eventually excreted in the urine; retention by the mesenteric lymph nodes appeared to account for the pigmentation observed in this tissue (Selim, 1982).

Toxicological studies

Special studies on carcinogenicity (See also long-term studies)

Mice

A carcinogenicity study of caramel colour IV was performed on $B_6C_3F_1$ mice using 5 groups of 50 male and 50 female mice in each group. Two groups of each sex served as contols and the 3 treatment groups received caramel colour IV in the drinking water at dose levels of 2.5, 5.0, or 10.0 g/kg b.w./day for 104 weeks. All animals were observed twice daily and palpated for tissue masses weekly from week 27. Body weights and food intake were recorded weekly, and fluid intake was measured on the first, third, fifth, and seventh days of the first 12 weeks and over a 48-hour period each week thereafter.

A complete necropsy was performed on all animals dying on test or sacrificed in a moribund condition, and on all survivors at termination. Histology (haematoxylin and eosin staining) was performed on all gross lesions and on the following organs/tissues of all animals: adrenals, brain (three sections), bone, bone marrow, epididymis, oesophagus, eyes, gall bladder, heart, intestine (duodenum, ileum, caecum and colon), kidneys, liver, lymph nodes (cervical, mesenteric, and thoracic), mammary glands, ovaries, pancreas, parathyroid, pituitary, prostate, salivary gland, sciatic nerve, seminal vesicle, skin, spinal cord, spleen, stomach, testes, thymus, thyroids, trachea, urinary bladder, and uterus.

Four animals were found in the wrong cages during week 15 and had probably been misplaced 9 weeks earlier. These animals were removed from the study during week 31 and were not included in subsequent examinations.

Although during the course of this study there were sporadic significant differences in mean body weights of treated male and female groups compared to controls, caramel-colour feeding did not have consistent effects on body weights or body-weight gains in either sex. Males of the 10 g/kg b.w.-dose group had lower mean food consumption than the control groups for 67 of the 104 weeks and males receiving 2.5 g/kg b.w. had lower mean food consumption for 21 of the 104 weeks. There were no consistent differences in food intake of males at the 5 g/kg b.w. dose level nor of any of the females. Decreased fluid intake was noted, particularly in males at the higher dose levels; mean fluid intakes of treated females were usually equal to or only slightly lower than controls. There were no treatment-related differences in survival rates, which at termination were 65-75% for the male groups and 67-79% for the female groups. At necropsy, there were dose- and/or treatment-related effects on the gastrointestinal tract and mesenteric lymph nodes, including dark gastrointestinal contents, staining of the mucosa, and diffusely red and congested mesenteric lymph nodes. These changes were not considered to be toxicologically important and there were no other changes of toxicological significance. There was no evidence of treatment-related neoplastic lesions in any organs (MacKenzie, 1985b).

Rats

A long-term toxicity and carcinogenicity study of caramel colour IV was conducted on F344 rats. In the carcinogenicity portion of the study, 5 groups of 50 male and 50 female weanling rats were selected randomly; 2 groups of each sex served as controls and the 3 treatment groups received caramel colour IV in drinking water at dose levels of 2.5, 5.0, or 10 g/kg b.w./day for 24 months. In the chronic toxicity portion of the study, groups of 25 male and 25 female rats received caramel colour IV in drinking water at doses of 0, 2.5, 5.0, 7.5, or 10 g/kg b.w./day for 1 year. The following parameters were monitored: mortality, clinical observations including ophthalmic changes, body weight, and food and fluid consumption. In the chronic toxicity portion clinical chemistry and haematological studies were done at intervals of 6 and 12 months and necropsies were performed at 12 months.

In the carcinogenicity portion of the study, a complete necropsy was performed on all animals dying on test or sacrificed in moribund condition, and on all survivors at termination. Histology (haematoxylin and eosin staining) was performed on all gross lesions and on the following organs/tissues of all animals: adrenals, brain (three sections), bone, bone marrow, epididymis, oesophagus, eyes, heart, intestine (duodenum, ileum, caecum, and colon), kidneys, liver, lung, lymph nodes (cervical, mesenteric, and thoracic), mammary gland, ovaries, pancreas, parathyroid, pituitary, prostate, salivary gland, sciatic nerve, seminal vesicle, skin, spinal cord, spleen, stomach, testes, thymus, thyroid, trachea, urinary bladder, and uterus.

The feeding of caramel colour IV did not affect survival in either the chronic toxicity or carcinogenicity sections of this study and, other than dark-stained and soft faeces, there were no treatment-related antemortem observations. During the course of both studies body weights were reduced for both males and females at the 5 and 10 g/kg b.w. dose levels. These effects on body weights were correlated with reduced water and food consumption at these dose levels and reflect the reduced palatability of the drinking fluid.

Clinical chemistry and haematological studies at 6 and 12 months in the chronic toxicity study did not reveal changes of toxicological concern. At 6 months, serum concentrations of blood urea nitrogen and creatinine were reduced in male groups treated with 5.0, 7.5, and 10 g/kg b.w. and in female groups treated with 7.5 and 10 g/kg b.w. caramel colour. Similar changes were noted at 12 months. Creatinine levels were within the normal range for the F344 rat, whereas blood urea nitrogen levels were slightly outside this range. Decreased levels of serum total protein, albumin, and globulin were also noted at the 6-month sampling, particularly in male rats. These changes, which were less marked at 12 months, were not accompanied by any pathology in the liver or kidneys and were not considered to be of toxicological importance.

The urinalysis studies revealed generally reduced urine volume and increased specific gravity in both sexes.

At necropsy the changes noted were characteristic of the feeding of high levels of caramel colour, which consisted primarily of brown staining of the gastrointestinal tract and mesenteric lymph nodes and caecal enlargement. Increased kidney weights were noted in both sexes in animals fed caramel colour IV; no histologic alterations were present that could be associated with the increased renal weight, which the authors considered to be related to the water imbalance in these animals.

Microscopic examination of the tissues of these animals did not reveal specific toxicological changes. Pigmentation in the gastrointestinal tract and mesenteric lymph nodes was observed. There was no evidence of reactive hyperplasia to the pigment in the mesenteric lymph nodes of the gastrointestinal tract.

In the chronic portion of the toxicity study, although statistically significant changes were noted in some parameters, they were not considered to be toxicologically important. The highest dose fed, 10 g/kg b.w., was considered to be the no-adverse-effect level.

In the carcinogenicity portion of the study, the observations were generally similar to those in the chronic portion. Survival at 24 months ranged from 64-68% for males and 82-92% for females. Random variations in both benign and malignant neoplasms typical of the F344 strain and this age of animal were observed; however, there were no treatment-related differences. The authors concluded that the feeding of caramel colour IV at doses up to 10 g/kg b.w. for 24 months did not induce neoplastic changes or non-neoplastic changes of toxicological importance (MacKenzie, 1985a).

Special studies on mutagenicity

Caramel colour IV was evaluated for genetic acitivity in a series of in vitro microbial assays with and without metabolic activation. Salmonella typhimurium (strains TA1535, TA1537, and TA1538) and Saccharomyces cerevisiae were used. Caramel colour IV was not genetically active under the test conditions employed in this study (Brusick, 1974).

Caramel colour IV was evaluated for mutagenicity using the Ames Salmonella/microsome plate test and the Saccharomyces/microsome plate test. The samples tested were blends of 3 samples of caramel colour IV (low colour intensity) and 3 samples of caramel colour IV (high colour intensity). The Salmonella test organisms used were TA98, TA100, TA1535, TA1537, and TA1538. Tests were conducted directly or in the presence of liver microsomal enzyme preparations from Araclor-induced rats. Tests were conducted over a range of concentrations from 1-50 mg/plate. No signs of genetic activity were observed with any of the samples of caramel colour IV tested using either the Salmonella or Saccharamyces test organisms (Jagannath & Brusick, 1978a).

The clastogenicity of caramel colour IV (low colour intensity) was evaluated in an in vitro cytogenetic assay using cultured Chinese hamster ovary cells without metabolic activation. No increase in chromosome aberrations was observed at concentrations of test material between 5 μg/ml and 5 mg/ml. In the same test, sodium ascorbate produced a positive reponse at 2 mg/ml (Galloway & Brusick, 1981a).

The clastogenicity of caramel colour IV (high colour intensity) was evaluated in an in vitro cytogenetic assay using Chinese hamster ovary cells without metabolic activation. No increase in chromosome aberrations was observed at concentrations of test material between 5 μg/ml and 5 mg/ml. In the same test, sodium ascorbate produced a positive response at 2 mg/ml (Galloway & Brusick, 1981b).

Two samples of caramel colour IV of medium and high colour intensity were assayed for mutagenic activity in the Ames test using Salmonella strains TA98 and TA100 in the absence or presence of rat hepatic S-9 fraction. Neither sample showed mutagenic activity under the conditions of the test (Ashoor & Monte, 1983).

Thirteen commercial caramel colours (not identified) were examined for mutagenicity in the Ames test using Salmonella strains TA98 and TA100, with and without metabolic activation, and for DNA-damage effects on E. coli (Wild/pol A⁻; Wild/rec A⁻). None of the caramel-colour samples tested showed mutagenic activity in these tests (Kawana et al., 1980).

Five samples of commercial caramel colour were tested for mutagenic activity against Salmonella strains TA98 and TA100 in the Ames test. All samples were reported to show equivocal results with strain TA100, and 2 were equivocal with TA98 (Kawachi et al., 1980).

Five samples of caramel colour were tested in a chromosome aberration test in a cultured cell line of Chinese hamster lung fibroblasts, in the presence or absence of hepatic S-9 fraction. Ames tests were also conducted using Salmonella strains TA98, TA100, and

TA1537, with or without metabolic activation and with a 20-minute pre-incubation step. All samples of caramel colour were designated as positive in the Ames assay and in the chromosome aberration test (aberrations were detected in 20% of metaphase cells) (Ishidate & Yoshikawa, 1980).

The same series of 5 caramel-colour samples as used in the above studies were tested for mutagenicity in the Ames test using Salmonella strains TA98, TA100, TA1535, TA1537, and TA1538 and Saccharomyces strain D4. Assays were performed both in the presence and absence of hepatic microsomal S-9 fraction from Araclor-induced rats in the conventional plate assay and after a pre-incubation step in strains TA100 and TA1535. The results of the mutagenicity assays were negative under all the test conditions and over a concentration range of 1-50 mg/plate for strain TA100 and 1-20 mg/plate for the other strains (Jagannath & Brusick, 1978b).

Special studies on reproduction

Fifteen male and female Wistar rats were given 0 or 10% caramel colour IV solution as their sole fluid source until day 100 and were then mated. Animals of the F_1-generation (25 males and 25 females) were weaned and again given 0 or 10% caramel solution until day 100. There were no adverse effects with regard to the number of litters born and the number of pups/litter. No influence on haematology, growth, food consumption, gross pathology, or histopathology of the F_1-generation at 100 days of age was observed (Haldi & Wynn, 1951).

Six different samples of caramel colour IV (3 single-strength (SS) and 3 double-strength (DS) products), each containing a different level of 4-methylimidazole (between 200 and 850 ppm) were tested in a reproduction study in rats. Two control diets were used, an unmodified stock diet and the stock diet supplemented with starch and cellulose. The various test diets are shown in the following diagramme.

Composition of the test diet (in g)

Group	1	2	3	4	5	6	7	8	9	10	11	12
Caramel in %	5	10	15	10	10	2	4	6	4	4	0	0
stock diet	100	100	100	100	100	100	100	100	100	100	100	100
wheat starch	3.8	1.9	-	1.9	1.9	4.9	4.1	3.4	4.1	4.1	5.8	-
cellulose	4.2	2.2	-	2.2	2.2	5.4	4.6	3.8	4.6	4.6	6.2	-
SS[a] caramel colour IV + 202[b]	5.6	11.5	17.6									
SS caramel colour IV + 400				11.5								
SS caramel colour IV + 600					11.5							
DS[c] caramel colour IV + 350						2.3	4.5	6.8				
DS caramel colour IV + 639									4.5			
DS caramel colour IV + 852										4.5		

[a]SS=Single stength
[b]Quantity of 4-methylimidazole in ppm
[c]DS=Double strength

Twelve groups of 10 male and 20 female weanling Wistar rats were allocated to the above dietary groups and, at week 12, the rats were mated in sub-groups of 5 males and 10 females. After a 3-week mating period, the females were caged individually.

At weaning age of the F_1 generation, 40 males and 40 females were selected from as many different litters as possible within each caramel-colour group and continued on the same diet as the parent generation. Ten males and 10 females of each group were sacrificed

after one year (see Sinkledam et al., 1975, under short-term toxicity
studies); the remaining 30 males and 30 females of each group were fed
the test diets for 2 years (see Sinkeldam et al., 1976, under long-term
toxicity studies). After weaning the F_1 litters, the dams were
killed and the implantation sites were counted.

No consistent, dose-related effects on growth of the F_0
animals were noted. No adverse effects were seen on female fertility,
litter size, average weight and growth of the pups, or number of
implantation sites or sex ratio of the young. In one group (10% SS
caramel colour IV + 600 ppm 4-methylimidazole) there was a slight
increase in mortality at birth. No teratogenic effects were found (Til
& Spanjers, 1973).

A range-finding reproduction study was conducted on F344
rats. Five groups of 12 male and 12 female mature rats were given
caramel colour IV at concentrations of 0, 10, 15, 20, or 25% in
drinking water for 21 days prior to mating and throughout mating,
gestation, and lactation. Animals of the F_0 generation were killed
when animals from the F_1 generation were weaned. At weaning, 2
pups/sex/litter were randomly selected from litters that had a minimum
of 2 males and 2 females at day 21 and treatment of these animals was
continued for 13 weeks post-weaning. Haematology and clinical
chemistry were performed on days 45/46 and at termination. Complete
gross post-mortem examinations of both F_0 and F_1 animals were
performed at sacrifice. Selected organs (caecum, spleen, thymus,
liver, kidneys, heart, adrenals, and gonads from animals of the F_0
generation were weighed at autopsy.

Although the dose levels of caramel colour IV in this study
were very high (ranging from 8-28 g/kg b.w.), no specific toxic effects
were observed. All F_0-generation animals survived the duration of
the study, but 3 F_1-generation animals were killed accidentally while
sampling for clinical chemistry studies at day 45. In the 20- and
25%-dose groups of both generations there was a higher incidence of
soft stools than in the controls, and all animals of the F_0
generation showed slight to statistically significant, dose-related
decreases in body weight. Dose-related depression of body-weight gain
was also noted in animals of the F_1 generation.

Mating, pregnancy, and fertility rates were comparable for all groups, but the number of implantation sites and of pups alive at days 0, 4, and 21 of lactation in the 20%-dose grops was significantly lower than control values. Litter size was decreased at the 15-, 20- and 25%-dose levels. Pups in the 25%-dose group showed a markedly higher incidence of alopecia and arched spine than controls, and a generalized poor condition during the last 7 days of suckling. No significant haematological changes were observed at 45 or 90 days post-weaning, except that prothrombin time of the F_1 females in the 15 and 25% groups were significantly greater than controls at day 45. Blood urea nitrogen values were lower than controls at 45 and 90 days but other clinical chemistry values were normal.

At necropsy, dose-related increases in absolute and relative weights of the liver, kidneys, and caecum (full and empty) were observed from animals in the 15%- and higher-dose groups. The only gross treatment-related morphological changes reported were brown/black or green colouration of the contents and mucous membrane of the lower gut and mesenteric lymph nodes (Tierney, 1980).

Special studies on teratogenicity

Teratogenicity tests were performed with caramel colour IV on mice, rats, and rabbits. The doses employed were 0, 16, 74.3, 345, and 1600 mg/kg b.w. in all 3 species.

Mice

Caramel colour IV was administered by gavage to groups of 19-22 pregnant albino CD1 mice at the doses above, beginning on day 6 and continuing through day 15 of gestation. Body weights were recorded and all animals were observed daily for changes in appearance and behaviour. On day 17 all dams were subjected to Caesarean section under surgical anaesthesia and the numbers of implantation sites, resorption sites, live and dead foetuses, and body weights of live pups were recorded. All foetuses were examined grossly for external congenital abnormalities, one-third of the foetuses from each litter underwent visceral examination (Wilson technique), and the remaining two-thirds were cleared, stained with Alizarin Red, and examined for skeletal defects.

The number of abnormalities seen in either soft or skeletal tissues did not differ from the number occurring spontaneously in sham-treated controls (Morgareidge, 1974).

Rats

Caramel colour IV was administered by gavage to groups of 21-24 pregnant Wistar rats at the dose levels above, beginning on day 6 and continuing daily through day 15 of gestation. Body weights were recorded and all animals were observed daily for changes in appearance and behaviour. On day 20 dams were subjected to Caesarean section under surgical anaesthesia and the number of implantation sites, resorption sites, live and dead foetuses, and body weights of live pups were recorded. All foetuses were examined for gross, visceral, and skeletal abnormalities as in the mouse experiment.

No clearly discernible effects on nidation or on maternal or foetal survival were observed. The number of abnormalities seen in the test groups did not differ from the number occurring spontaneously in sham-treated controls (Morgareidge, 1974).

Rabbits

Caramel colour IV was administered by gavage to groups of 11-12 pregnant Dutch-belted female rabbits at the doses above beginning on day 6 and continuing daily through day 18 of pregnancy. Body weights were recorded and all animals were observed daily for changes in appearance and behaviour. The does were subjected to Caesarian section under surgical anaethesia on day 29 and the numbers of corpora lutea, implantation sites, resorption sites, and live and dead foetuses were recorded.

Body weights of the live pups were also recorded. All foetuses were examined grossly for the presence of external congenital abnormalities. The live foetuses from each litter were then placed in an incubator for 24 hours for evaluation of neonatoal survivial. Surviving pups were sacrificed and all pups examined for visceral abnormalities (by dissection). All foetuses were then cleared with potassium hydroxide, stained with Alizarin Red, and examined for skeletal defects.

No clearly discernible effects on nidation or on maternal or foetal survival were observed. The number of abnormalities seen in either soft or skeletal tissues of the test groups did not differ from the number occurring spontaneously in sham-treated controls (Morgareidge, 1974).

Special study on haematology
Rats

A study was conducted in rats to determine whether haematological changes were associated with the feeding of high dietary concentrations of caramel colour IV. Sixty-six male rats, mean initial body weights approximately 173 g, were fed caramel colour IV in the diet for 28 days. Sixteen rats were assigned to a control group and 10 rats/group were assigned to levels of 16 or 22% caramel colour IV (high colour intensity) and 34 or 47% caramel colour IV (low colour intensity). Caramel colour III was fed at a 4% level to a positive control group. The parameters observed were body weight, food consumption, haematology (total and differential leucocyte counts), and terminal necropsy with organ weights (caecum - full and empty). Rats were bled from the retro-orbital sinus on days 16, 9, and 2 before treatment and again on days 7, 14, and 28 during treatment.

Feeding of these high dietary levels of caramel colour IV led to reduced rates of body-weight gain. The animals fed caramel colour III gained weight at a rate similar to the controls. Animals receiving caramel colour III had progressively lower relative lymphocyte counts commencing 1 week after treatment and increasing in severity with duration of treatment. This pattern of decreased relative lymphocyte counts was not seen with either sample of caramel colour IV and no significant decreases in lymphocyte counts were observed at any dose level of this caramel colour. Caecal weights were increased in all treatment groups (BIBRA, 1978).

Acute toxicity
No information available.

Short-term studies

Mice

Groups of 10 $B_6C_3F_1$ mice of each sex were given concentrations of 0, 10, 15, 20, or 30% caramel colour IV in drinking water for 4 weeks. The intake of caramel colour IV expressed in g/kg b.w./day was more than twice the percentage concentration in the drinking water. At 28 days the body weights of male animals at the 30%-dose level were significantly decreased compared to the controls, and transient depressions of body weights were noted at the highest-dose level in females. No significicnat depressions in body-weight gain were noted at the lower-dose levels in either male or female mice. Fluid consumption was depressed throughout the study among all treatment goups. However, these changes were not consistent with time or dosage. There were no statistically significant differences in food consumption between treated animals and controls. At necropsy, the only treatment-related effect reported was enlargement of the caecum (Tierney, 1979).

Rats

Groups of 5 rats received either 10 or 20% caramel colour IV solution (equivalent to about 10 or 20 g/kg b.w./day) as their sole source of fluid for 127 days. Only dark faeces and very mild diarrhoea were noted. No adverse effects were noted regarding general health, body weight, food and fluid consumption, haematology, gross pathology, or histopathology (Haldi & Wynn, 1951).

Six groups of 5 male and 5 female weanling rats received 0 or 10% caramel colour IV solution as their sole fluid source for 100, 200, or 300 days. No adverse effects were noted regarding growth, food and fluid intake, haemotology, gross pathology, or histopathology (Haldi & Wynn, 1951).

Groups of 16 male and 16 female rats received either 0 or 10% caramel colour IV solution for 100 days and groups of 5 rats received 20% caramel solution for 100 days. At the lower test level, there were no observable abnormalities as regards growth, food consumption, haematology, gross pathology, or histopathology. Only growth and haematology were examined at the higher test level, and the results for both parameters were found to be normal (Haldi & Wynn, 1951).

Goups of 5 male and 5 female rats were given 1 ml/kg b.w. of concentrated caramel colour for 21 days. Some diarrhoea was induced in all animals, but no other abnormalities were noted. Gross pathology and histopathology revealed no significant changes due to administration of the test compound (Foote et al., 1958).

Three groups of 20 male and female rats received either 0 or 11-14 g/kg b.w. of caramel colour IV solutions for 100 days. Growth and food intake did not differ significantly between test and control animals. Gross pathology and histopathology showed no abnormal findings related to administration of the test compound (Haldi & Wynn, 1958).

Four groups of 10 male and 10 female Sprague-Dawley rats received 0, 0.1, 1.0 or 10% caramel colour IV in their diet for 12 weeks. No adverse effects were noted on growth, food consumption, urinalysis, haematology, gross pathology, or histopathology that were related to administration of the caramel colour (Prier, 1960).

Groups of 10 male and 10 female rats received 0, 5, or 10 g/kg caramel colour IV in their diet for 3 months. Weight gain was normal in all groups. Food consumption, haematology, and urinalysis were comparable in all groups. Gross pathology and histopathology showed no test-related adverse findings (Chacharonis, 1960).

Four groups of 10 male and 10 female Sprague-Dawley rats received 0, 5, 10, or 20% caramel colour IV (low colour intensity) or caramel colour IV (high colour intensity) in their diet for 90 days. In addition, a paired-feeding study involving 5 male rats in 2 groups was run for 23 days (one sample was at the 20% level), and there were no differences in the rate of growth. The only effects attributable to treatment were a mild depression in growth of male rats at the 10 and 20% levels due to unpalatability of the test diet. No other adverse findings were noted in growth, behaviour, mortality, haematology, urinalysis, gross pathology, organ weights, or histopathology (Kay and Calandra, 1962a; 1962b).

Four groups of 10 male and 10 female Sprague-Dawley rats received either 0 or 10% of 3 different caramel colours (one of which was a sample of caramel colour IV) in their diet for 90 days. Weight gains showed a slight reduction compared with controls but food consumption was normal for all groups. No abnormalities were noted with regard to haematology, urinalysis, gross pathology, or histopathology (Chacharonis, 1963).

Four groups of 15 male and 15 female rats received 0, 5, 10, or 20% caramel colour IV in their diet for 90 days. No adverse effects were noted on appearance, behaviour, survival, body weights, food intake, haematology, blood chemistry, urinalysis, organ weights, gross pathology, or histopathology (Oser, 1963).

Four groups of 10 male and 10 female rats received 0, 0.015, 0.3, or 3.0% caramel colour IV in their diet for 90 days. No differences between test and control animals were noted regarding body weight, food consumption, haematology, urinalysis, gross pathology, or histopathology (Nees, 1964).

Four groups of rats received 0, 4, 8, or 16% caramel colour IV in their diet for 3 months. No convulsions or other behavioural abnormalities or signs of neurological damage were seen. No macroscopic or microscopic pathological abnormalities were found in the central nervous system (Sharratt, 1971).

Three groups of 20 female Wistar rats received stock diet to which 0, 10, or 20% caramel colour IV containing 202 ppm 4-methylimidazole was added. During the second week of the experiment these levels of caramel colour IV were increased to 15 and 25% and in week 7, the levels were increaseed to 25 and 30% caramel colour IV. The diets were administered for 16 weeks followed by a 4-week recovery period.

Food consumption and growth of the test animals were comparable with the controls. Leucocyte counts, collected after 4, 8, 12, and 16 weeks, did not show statistically significant differences among the groups. The relative weights of the caecum, both filled and empty, were distinctly increased after 4 weeks of feeding caramel colour IV. After the recovery period of 4 weeks the increases had disappeared. The relative weight of the thymus was not affected.

Gross examination at autopsy after 16 weeks of feeding caramel clour IV revealed a dose-related, brown-greenish discolouration of the mesenteric lymph nodes in all test animals of the highest-dose level. After the recovery period of 4 weeks the colour change was less, but still visible. Microscopically, the lymph nodes of the test rats showed pigment accumulation which was not noticeably diminished after withdrawal of the caramel for 4 weeks. These results failed to confirm the decreased leucocyte counts that were observed in females fed 5 to 15% caramel colour IV in a 1-year feeding study (Sinkeldam & van der Heyden, 1975).

A 10-week feeding study with 18 groups of 10 male and 10 female Wistar rats was carried out with 3 caramel colours, including caramel colour IV (low colour intensity) and caramel colour IV (high colour intensity). The dietary concentrations were 0, 1.25, 2.5, 5, 10, and 15% caramel colour IV (low colour intensity) and 0, 0.5, 1, 2, 4, and 6% caramel colour IV (high colour intensity).

Both caramel colours caused loose stools at the highest-dose levels, although body-weight gains were not affected. Leucocyte (lymphocyte) counts were not affected in the animals fed samples of caramel colour IV at any of the dose levels. Only slight indications of caecal enlargement were observed. Minimal amounts of pigment were observed in the mesenteric lymph nodes of rats fed 2.5% and higher of the low colour intensity sample, and 1% and higher of the high colour intensity sample (Sinkledam and van der Heyden, 1976).

In a 10-week feeding study, 17 groups of 15 male and 15 female Sprague-Dawley rats were given various caramel colours that included caramel colour IV (low colour intensity) and caramel colour IV (high colour intensity). The dose levels fed were 1.25, 2.5, 5, 10, and 15% for caramel colour IV (low colour intensity) and 0.5, 1, 2, 4, and 6% caramel colour IV (high colour intensity). Two control groups were used.

In the animals fed caramel colour IV (low colour intensity) at the 15% level, the faeces became soft within 2 weeks. The water content of the faeces from these animals was higher than that of the controls, as was the water content of faeces from rats fed 6% caramel colour IV (high colour intensity). Body-weight gains were slightly decreased in male, but not female, rats fed both samples of caramel colour.

There were no significant changes in total white cell or lymphocyte counts in animals fed either sample of caramel colour IV. No macroscopic or microscopic evidence of abnormal pigmentation of the mesenteric lymph nodes was found in any group of either sample of caramel colour IV. Caecal weights were generally increased in all test groups fed both samples of caramel colour IV (Procter et al., 1976).

Groups of weanling Wistar rats were given caramel colour IV (low colour intensity) or caramel colour IV (high colour intensity) at concentrations of 0, 0.5, 1.0, 2.0. 4.0, or 16.0% in the diet for 10 weeks. Each group contained 15 male or 15 female rats except the control group (60 animals of each sex) and the 16%-dose group (10 animals of each sex). Food intake and growth were recorded and haematological studies were carried out. Lymph nodes, thymus, spleen, and caecum were examined histologically for distribution of pigment.

An additional 3 groups of 10 rats of each sex were given basal diet or diet containing 16% each of the caramel colours for 10 weeks. At the end of this period all the rats received basal diets. Haematological studies were performed on these animals at 3, 7, 14, and 28 days. Five rats of each sex were killed after 7 days and the remainder after 28 days of feeding basal diet (recovery experiment).

Decreased body-weight gains were noted in animals of both sexes fed 16% caramel colour IV (high colour intensity). No decreases were observed in the groups fed caramel colour IV (low colour intensity). Food intake was not consistently altered in any of the groups fed either sample of caramel colour IV. Occasionally, values for total leucocyte counts were significantly higher or lower than controls in the groups fed both samples of caramel colour IV; however, these changes were not consistent in direction and they were not dose-related. There were no consistent or dose-related differences in lymphocyte counts between the groups fed caramel colour IV and the controls. Liver weights were significantly increased in the group fed 16% caramel colour IV (high colour intensity). Increased relative kidney weights were observed in the groups fed 2, 4, and 16% caramel colour IV (high colour intensity) and 16% caramel colour IV (low colour intensity), although no histological changes were observed. Increased caecal weights were seen only at the 16% feeding level for both caramel colours. At necropsy pigmentation of the lymph nodes was seen at the 16% feeding level of both caramel colours. Microscopically, pigmentation was observed in the mesenteric lymph nodes in the males and females in the groups fed 16% caramel colour IV (high colour intensity). Relative weights of the liver and kidneys and caecal weights returned to normal during the recovery period (BIBRA, 1977).

Five groups of 30 male and 30 female weanling F344 rats were given caramel colour IV in drinking water at concentrations which provided intakes of 0, 15, 20, 25, or 30 g/kg b.w./day for 13 weeks. After 43 days, 10 animals of each sex from each group were randomly selected for collection of data on urinalysis, haematology, and clinical chemistry, followed by sacrifice and necropsy; similar examinations were performed on survivors at termination. At interim and terminal sacrifice, a detailed necropsy was performed on all animals and a complete histopathological examination was performed on controls and animals in the top-dose group.

Throughout the study, rats given caramel colour IV produced dark-coloured, soft or sticky, liquid and/or odorous, faeces which stained and caused alopecia of the perianal area, most noticeably at the higher-dose levels. Dose-related decreases in food intake, water consumption (after correction for caramel content), and body-weight gains were observed and were attributed to the poor palatability of the drinking solutions.

Caramel colour IV at the dose levels tested did not signif-icantly affect haematological values at interim or terminal examina-tions. All treatment groups of both sexes had significantly-reduced blood urea nitrogen and alkaline phosphatase levels at both 45 and 90 days. Total serum protein values of both sexes in the treatment groups were lower than controls at 90, but not at 45, days. These effects may be due to reduced food intake and growth retardation.

Treated rats had reduced urine volume and increased urine specific gravity, protein, ketones, and acidity, which were associated with decreased water consumption. Increased kidney weights were observed at necropsy. Treatment-related decreases in thymus and spleen weights and increased caecal size, with staining of the mucosa of the caecum and colon, were noted. Accumulation of yellowish-tan pigment occurred in macrophages of the mesenteric lymph nodes. No treatment-related histopathological changes were observed in any organs at the highest-dose level (Heidt & Rao, 1980).

A 1-year rat toxicity study was conducted as a continuation of the reproduction study described earlier (Til & Spanjers, 1973). It involved the interim sacrifice of one-fourth of the animals utilized in the 2-year toxicity study described below.

Wistar rats (10 males or 10 females in each group) were selected from the first litter of parents that were given diets containing 6 samples of caramel colour IV, 3 of which were low colour intensity and 3 of which were high colour intensity. The dose levels employed were 5, 10, and 15% caramel colour IV (low colour intensity) and 2, 4, and 6% caramel colour IV (high colour intensity). Two additional samples of caramel colour IV (low colour intensity) were tested at 10% and 2 additional samples of caramel colour IV (high colour intensity) were tested at 4% in the diet. The animals were sacrificed after a feeding period of 1 year.

No adverse effects on behaviour, growth, food intake, mortality, liver and kidney function tests, urine composition, or organ weights were observed. Haematological indices showed a slight dose-related decrease in total leucocyte counts in females fed one sample of caramel colour IV (low colour intensity) which contained 202 mg/kg methylimidazole. In males, no effect was noted and in a subsequent experiment in which the same sample of caramel colour IV was fed to female rats at levels as high as 15 and 30%, no indications of decreased leucocyte counts were observed after 4, 8 , 12, or 16 weeks.

The only finding attributed to the feeding of caramel colour
IV consisted of increased accumulation of a yellow-brown pigment and
pigment-laden macrophages in the mesenteric lymph nodes of males and
females in all caramel colour IV groups. Inflammatory or degenerative
changes of the lymphoid tissue were not found (Sinkeldam et al., 1975).

Dogs

Four groups of 3 male and 3 female adult beagle dogs
received 0, 6, 12.5, or 25% caramel colour IV in their diet 5 days per
week for 90 days. No significant adverse effects on growth, behaviour,
food consumption, mortality, liver function, kidney function,
haematology, urinalysis, gross pathology, or histopathology were noted
(Kay and Calandra, 1962c).

Long-term toxicity studies

Rats

Six samples of caramel colour were tested in a long-term
toxicity study, 3 of which were caramel colour IV (low colour
intensity) and 3 of which were caramel colour IV (high colour
intensity). Each group consisted of 40 male or 40 female weanling
Wistar rats, except the control group which had double this number of
animals. Animals were selected from the first litter of parents fed
diets containing the various caramel colours from weaning age (see
reproduction studies, Til & Spanjers, 1973). The dose levels tested
were 5, 10, or 15% caramel colour IV (low colour intensity) and 2, 4,
or 6% caramel colour IV (high colour intensity). Two additional
samples of caramel colour IV (low colour intensity) were tested at 10%
and 2 additional samples of caramel colour IV (high colour intensity)
were tested at 4% in the stock diet.

Observations were made of general appearance, behaviour,
mortality, growth, food intake, haematological factors, and clinical
chemistry of blood and urine. After about 14 months mortality
attributed to intercurrent disease was observed in both control and
treated groups. Approximately three-quarters of the animals died or
were killed before the experiment was terminated at week 104. Organs
of animals that died or were killed were weighed and extensive
histopathological examinations were carried out. However, one-third of

the animals could not be examined histopathologically due to autolysis. At week 104 organs were weighed and extensive histopathological examinations were carried out on all surviving rats of all groups.

No clinical changes were observed in this study except for slightly decreased haemoglobin and haematocrit values at weeks 78 and 98 in males fed caramel colour IV (high colour intensity). Leucocyte counts were decreased in females fed 10 and 15% of one sample of caramel colour IV (low colour intensity) at weeks 13 and 52, but these changes were not consistent and decreases in lymphocyte counts were not reported. At autopsy, an increased incidence of greenish discoloured mesenteric lymph nodes was observed in most groups fed high levels of caramel colour. Microscopically, an increased pigment-phagocytosis in the mesenteric lymph nodes in all test groups (except the group fed 2% caramel colour IV (high colour intensity)) was observed. No evidence of any other adverse structural or cellular alteration was found. Gross and microscopic examination of the other organs did not reveal any pathological changes attributable to the ingestion of caramel colour IV. An increase in the incidence of neoplastic lesions in the different groups was not found (Sinkeldam et al., 1976).

A long-term toxicity and carcinogenicity study of caramel colour IV was conducted using F344 rats (MacKenzie, 1985a). Details are given above (see special studies on carcinogenicity).

Observations in man

Tolerance studies of caramel colour IV (low colour intensity) and caramel colour IV (high colour intensity) were conducted in human volunteers. The subjects, 10 men and 10 women, ingested caramel colour once daily in simulated soft drinks over 3 test periods of 21 days each separated by 7-day rest intervals. The test doses were 6 g/day during the first test period, 12 g/day during the second period, and 18 g/day during the third test period.

Haematological, clinical chemical, and routine urinary parameters were studied at the beginning of each ingestion period, after 10 days of ingestion, and at the end of each ingestion period. Most individual values for haemoglobin, haemotocrit, RBC, corrected sedimentation rate, WBC, and differentials (neutrophils, basophils, eosinophils, monocytes, and lymphocytes) were found to be within normal limits. There were a few instances of values outside the normal range, indicating mild neutropenia and mild lymphocytosis, but these were not consistent and were unrelated to the ingestion of caramel colour IV. On the other hand, caramel ingestion was associated with an increased frequency of bowel movements and softening or increased liquidity of faeces (Marier et al., 1977a; 1977b).

Comments

The pigmentation of mesenteric lymph nodes and caecal enlargement were considered to be non-specific effects that are of no toxicological significance. The carcinogenicity studies required by the twenty-fourth meeting of the Committee (Annex 1, reference 53) have been conducted in rats and mice, and no treatment-related neoplastic changes were observed. The material used in these studies conformed to recent specifications. The committee based its evaluation on the no-effect level in these long-term/carcinogenicity studies, to which (in view of the ancillary human data in which no adverse effects other than laxation were observed) a safety factor of 50 was applied.

EVALUATION
Level causing no toxicological effect
Mouse: 10 g/kg b.w./day in drinking water
Rat: 10 g/kg b.w./day in drinking water
Man: No adverse effects other than laxation at levels up
         to 18 g/day.

Estimate of acceptable daily intake for man
0-200 mg/kg b.w. (0-150 mg/kg b.w. on a solids basis).

REFERENCES

Ashoor, S.H. & Monte, W.C. (1983). Mutagenicity of commercial caramels. Cancer Letters 18, 187-190.

BIBRA (1977). The short-term (10 week) feeding study with three caramels in rats. Unpublished report No. 177/2/77 from the British Industrial Biological Research Association, Carshalton, Surrey, England.

BIBRA (1978). The evaluation of the effect of sulfite ammonia caramels on the lymphocytes of the rat. Unpublished report No. 238/2/78 from the British Industrial Biological Research Association, Carshalton, Surrey, England.

Brusick, D. (1974). Mutagenic evaluation of compound FDA 71-83, caramel. Unpublished report No. 2468 from Litton Bionetics Inc., Kensington, MD, USA.

Chacharonis, P. (1960). Acute and chronic toxicity studies on caramel colours A and B. Unpublished report No. S.A. 54219 from Scientific Associates Inc., St. Louis, MO, USA.

Chacharonis, P. (1963). Acute oral toxicity study in rats on caramel colorings 25A-1, 30B-0, and 30F-1. Unpublished report No. S.A. 79105 from Scientific Associates Inc., St. Louis, MO, USA.

Foote, W.L., Robinson, R.F., & Davidson, R.S. (1958). Toxicity of caramel color products. Unpublished report of Battelle Memorial Institute, Columbus, OH, USA.

Galloway, S.M. & Brusick, D.J. (1981a). Mutagenicity evaluation of ETA-36-1 in an in vitro cytogenetic assay measuring chromosome aberration frequencies in Chinese hamster ovary (CHO) cells. Unpublished report No. 20990 from Litton Bionetics Inc., Kensington, MD, USA. Submitted to WHO by International Technical Caramel Association.

Galloway, S.M. & Brusick, D.J. (1981b). Mutagenicity evaluation of ETA-63-1S in an in vitro cytogenetic assay measuring chromosome aberration frequencies in Chinese hamster ovary (CHO) cells. Part II. Unpublished report No. 20990 from Litton Bionetics Inc., Kensington, MD, USA. Submitted to WHO by International Technical Caramel Association.

Haldi, J. & Wynn, W. (1951). A study to determine whether or not caramel has any harmful physiological effect. I. Unpublished report from Emory University, Atlanta, GA, USA.

Haldi, J. & Wynn, W. (1958). A study to determine whether or not caramel has any harmful physiological effect. II. Unpublished report from Emory University, Atlanta, GA, USA.

Heidt, M. & Rao, G.N. (1980). 90-day subacute toxicity study of caramel color (sulfite ammonia process) type SAC2 in rats. Unpublished report No. 79028 from Raltech Scientific Services, Inc., Madison, WI, USA. Submitted to WHO by International Technical Caramel Association.

Ishidate, M. Jr. & Yoshikawa, K. (1980). Chromosome aberration tests with Chinese hamster cells in vitro with and without metabolic activation – a comparative study on mutagens and carcinogens. Arch. Toxicol. Suppl. 4, 41-44.

Jagannath, D.R. & Brusick, D. (1978a). Mutagenicity evaluation of ETA 4-10, ETA 4-11, ETA 4-15 in the Ames Salmonella/microsome plate test. Unpublished report No. 20838 from Litton Bionetics Inc., Kensington, MD, USA. Submitted to WHO by International Technical Caramel Association.

Jagannath, D.R. & Brusick, D. (1978b). Mutagenicity evaluation of caramel colors ETA 4-8, ETA 4-9, ETA 4-12, ETA 4-13, ETA 4-14 in the Ames Salmonella/microsome plate test. Unpublished report No. 20838 from Litton Bionetics Inc., Kensington, MD, USA. Submitted to WHO by International Technical Caramel Association.

Kawachi, T., Yahagi, T., Kada, T., Tazima, Y., Ishidate, M., Sasaki, M., & Sugiyama, T. (1980). Cooperative programme on short-term assays for carcinogenicity in Japan. In Molecular and Cellular Aspects of Carcinogen Screening Tests, IARC Scientific Publication No. 27, 323-330. International Agency for Research on Cancer, Lyon, France.

Kawana, K., Akema, R., Nakaoka, T., Ikeda, H., Takimoto, T., & Kawauchi, S. (1980). Studies on mutagenicities of natural food additives. II. Mutagenicities and antibacterial activities of carmels. Eisei Kagaku 26, 259-263.

Kay, J.H. & Calandra, J.C. (1962a). Subacute oral toxicity of caramel colorings (coded Sample A) – albino rats. Unpublished report from Industrial Bio-test Laboratories, Inc., Northbrook, IL, USA.

Kay, J.H. & Calandra, J.C. (1962b). Ninety-day subacute oral toxicity of caramel coloring (coded Sample B) - albino rats. Unpublished report from Industrial Bio-test Laboratories, Inc., Northbrook, IL, USA.

Kay, J.H. & Calandra, J.C. (1962c). Subacute oral toxicity of caramel coloring (coded Sample A) - dogs. Unpublished report from Industrial Bio-test Laboratories, Inc., Northbrook, IL, USA.

MacKenzie, K.M. (1985a). Carcinogenicity and chronic toxicity study of caramel color (sulfite ammonia process) type SAC2 in rats (Chronic toxicity portion, Vols. I-V; Carcinogenicity portion, Vols. I-VIII). Unpublished study No. 81184 from Hazleton Laboratories America, Inc., Madison, WI, USA. Submitted to WHO by International Technical Caramel Association.

MacKenzie, K.M. (1985b). Carcinogenicity study of caramel color (sulfite ammonia process) type SAC2 in mice. Vols. I-V1. Unpublished study No. 81185 from Hazleton Laboratories America, Inc., Madison, WI, USA. Submitted to WHO by International Technical Caramel Association.

Marier, G., Berry, G., & Orr, J.M. (1977a) Tolerance study of single strength ammonia sulfite caramel in human volunteers. Unpublished project No. 5612, report No. 2, from Bio-Research Laboratories Ltd., Pointe Claire, Quebec, Canada.

Marier, G., Berry, G., & Orr, J.M. (1977b) Tolerance study of double strength ammonia sulfite caramel in human volunteers. Unpublished project No. 5711, report No. 2, from Bio-Research Laboratories Ltd., Pointe Claire, Quebec, Canada.

Morgareidge, K. (1974). Teratologic evaluation of FDA 71-83 (caramel, beverage) in mice, rats and rabbits. Unpublished report No. PB 234-867 from Food and Drug Research Laboratories Inc., Waverly, NY, USA.

Nees, P.O. (1964). Toxicological feeding study of caramel E. Unpublished report from Wisconsin Alumni Research Foundation, Madison, WI, USA.

Oser, B.L. (1963). Toxicological feeding study of "acid-proof" caramel. Unpublished report No. 83911 from Food and Drug Research Laboratories, Inc., Waverly, NY, USA.

Prier, R.F. (1960). The toxicity of double strength acid proof caramel in rats - 12 week feeding test. Unpublished report No. 9070599 from Wisconsin Alumni Research Foundation, Madison, WI, USA.

Procter, B.G., Berry, G., & Chappel, C.I. (1976). A toxicological evaluation of various caramels fed to albino rats. Unpublished report No. 4244 from Bio-Research Laboratories Ltd., Pointe Claire, Quebec, Canada.

Selim, S. (1982). Single and repeat oral dose pharmacokinetic and distribution studies of caramel color concentrate in the rat. Unpublished study No. IT-59r from Primate Research Institute, Holloman Air Force Base, NM, USA. Submitted to WHO by International Technical Caramel Association.

Sharratt, M. (1971). Unpublished report.

Sinkeldam, E.J. & van der Heyden, C.A. (1975). Short-term feeding study with caramel SS 202 in albino rats. Unpublished report No. R 4777 from Centraal Instituut voor Voedingsonderzoek (CIVO/TNO), Zeist, The Netherlands.

Sinkeldam, E.J. & van der Heyden, C.A. (1976). Short-term (10 week) feeding study in rats with three different ammonia caramels. Unpublished report No. R 5120 from Centraal Instituut voor Voedingsonderzoek (CIVO/TNO), Zeist, The Netherlands.

Sinkeldam, E.J., van der Heyden, C.A., & Beems, R.B. (1976). Chronic (two year) feeding study in rats with six different ammonia caramels. Unpublished report No. R 4961 from Centraal Instituut voor Voedingsonderzoek (CIVO/TNO), Zeist, The Netherlands.

Sinkeldam, E.J., Willems, M.I, & van der Heyden, C.A. (1975). One year feeding study in rats with six different ammonia caramels. Unpublished report No. R 4767 from Centraal Instituut voor Voedingsonderzoek (CIVO/TNO), Zeist, The Netherlands.

Tierney, W.J. (1979). A four-week dose range-finding study of caramel colour no. 3, sample 3-1, in mice. Unpublished project No. 79-2380 from Bio/dynamics Inc., East Millstone, NJ, USA. Submitted to WHO by International Technical Caramel Associaton.

Tierney, W.J. (1980). An in-utero range-finding study of caramel colour no. 3, sample 3-1, in rats. Unpublished project No. 79-2405 from Bio/ynamics Inc., East Millstone, NJ, USA. Submitted to WHO by International Technical Caramel Associaton.

Til, H.P. & Spanjers, M. (1973). Reproduction study in rats with six different ammonia caramels. Unpublished report No. R 4068 from Centraal Instituut voor Voedingsonderzoek (CIVO/TNO), Zeist, The Netherlands.

FAST GREEN FCF

EXPLANATION

Fast Green FCF was evaluated at the twenty-fifth meeting of
the Committee (Annex 1, reference 56) when inadequacies were identified
in earlier long-term feeding studies in rats and mice. The previously-
allocated ADI was converted to a temporary ADI of 12.5 mg/kg b.w. pending the results of adequate long-term feeding studies and multigeneration reproduction/teratogenicity studies.

Since the previous evaluation, additional data have become
available and are summarised and discussed in the following monograph.
The previously-published monograph has been expanded and is reproduced
in its entirety below.

BIOLOGICAL DATA

Biochemical aspects

Rats and dogs were given orally 200 mg of the colour. In
the rats the urine and faeces were collected for 36 hours. In the
dogs, a bile fistula was made for bile analysis. Almost all the
administered colour was excreted unchanged in the faeces of rats. No
colour was found in the urine. In the bile of the dogs, the amount of
colour never exceeded 5% of the given dose. After feeding, the colour
was found in the bile of rats and rabbits, but not in their urine. It
was concluded that the quantity found in the bile provides a reasonable
estimate of the amount absorbed from the gastrointestinal tract (Hess &
Fitzhugh, 1953; 1954; 1955).

Following i.v. injection in rats, over 90% of the colour was
excreted in the bile within 4 hours (Iga et al., 1971).

Fast Green FCF was found to have a high binding affinity for plasma protein (Gangolli et al., 1967; 1972; Iga et al., 1971).

Toxicological studies

Special studies on carcinogenicity (see also long-term studies)

Mice

Groups of 60 (120 controls) male and female Charles-River CD-1 mice were fed diets containing 0, 0.5, 1.5, or 5.0% Fast Green FCF from 43 days of age for approximately 24 months. Ten animals/sex/group were subjected to haematological examination at 3, 6, 12, and 18 months. All animals dying or killed in a moribund condition and all survivors to termination were subjected to detailed post-mortem examination. The following tissues were examined histologically from all survivors from the control and 5%-dose groups as well as all animals dying or killed in extremis from these groups: adrenals, aorta, bone and marrow (femur), brain (3 sections), eyes (with optic nerve), gall bladder, gastrointestinal tract (oesophagus, stomach, duodenum, ileum, caecum, colon), heart, kidneys, liver, lung, lymph nodes (mesenteric and mediastinal), mammary gland, nerve (sciatic), ovaries, pancreas, pituitary, prostate, salivary gland, seminal vesicles, skeletal muscle, skin, spinal cord, spleen, testes with epididymides, thymus, thyroid/parathyroid, trachea, urinary bladder, uterus, and gross lesions/tissue masses. In addition, gross changes/tissue masses were examined histologically from all animals in the lower-dose groups.

No treatment-related effects on mortality were observed. The mean body weights of females in the 5%-dose group were consistently lower than controls and the mean body weights of males in the 5%-dose group were lower than controls at weeks 52 and 78. No other consistent differences in body weight were noted. Slight reductions in haemoglobin, haematocrit, and erythrocyte counts were noted in the high-dose males at 18 months but no other consistent or dose-related haematological changes were observed. Histological examination did not reveal any treatment-related lesions and the incidence, origins, and histology of benign and malignant neoplasms did not differ significantly between controls and treated animals (Hogan & Knezevich, 1981).

Rats

Eighteen weanling Osborne-Mendel rats of both sexes received weekly s.c. injections of approximately 30 mg (1 ml of a 3% aqueous solution) of Fast Green FCF for 94-99 weeks. Subcutaneous fibrosarcomas appeared at the site of injection in 15 animals (Nelson & Hagan, 1953; Hansen et al., 1966).

Two groups of 16 female rats (control groups of 10 rats) were given s.c. injections of 0.5 ml of a 3% or 6% solution (the rats received with each injection 15 or 30 mg, respectively). The colour used in the experiment was certified as 92% pure and was supplied as the disodium sulfonate salt. The 10 control rats were given distilled water injections. At first, injections of 6% were given 3 times a week; after 17 weeks it became necessary to reduce the dose to 3%. Thereafter, both groups were given injections of 3% twice weekly for 9 weeks. The rest of the time, 22 weeks, both groups were injected usually once a week, while occasionally 2 injections were tolerated. Growth inhibition was observed. Thirteen out of 16 animals receiving 6% of the colour had fibrosarcomas. The animals given 3% also showed fibrosarcomas (10 out of 12). The controls did not show neoplastic tissue at the site of injection (Hasselbach & O'Gara, 1960).

Subcutaneous injection of 1 ml of an 0.8% solution twice weekly produced histological changes suggestive of subsequent sarcoma formation unassociated with chemical carcinogenic potential (Grasso & Golberg, 1966).

No tumours were produced in 11 hamsters injected with 1 mg of the dye in 0.1 ml water (Price et al., 1978).

A carcinogenicity study with an in utero phase was carried out in Charles-River albino rats. Groups of 60 (120 controls) male and female rats of the F_0 generation were fed diets containing 0, 1.25, 2.5, or 5.0% Fast Green FCF for 2 months prior to mating and throughout gestation and lactation. Following the reproductive phase, a maximum of 4 animals of each sex/litter were randomly selected from the F_1

generation for the long-term carcinogenicity study. Groups of 70 animals of each sex/group were given Fast Green FCF in the diet at the same concentrations as the parent generation. An interim kill of 10 animals of each sex per group was carried out after 12 months; the remaining animals continued to receive the test diets for 29 months (males) or 31 months (females). Haematology, clinical chemistry tests, and urinalysis were performed on 10 rats of each sex/group at 3, 6, 12, 18, and 24 months. Gross autopsies were performed on all animals that died on test or were killed in a moribund condition and on all F_1 generation animals at interim and terminal sacrifice. The following tissues were examined histologically from all animals killed at interim sacrifice and all survivors from the control and 5%-dose groups, as well as all animals dying or killed in extremis from these groups: adrenals, aorta, bone and marrow (femur), brain (3 sections), eyes (with optic nerve), gastrointestinal tract (oesophagus, stomach, duodenum, ileum, caecum, colon), heart, kidneys, liver, lung, lymph nodes (mesenteric, mediastinal) mammary gland, nerve (sciatic), ovaries, pancreas, pituitary, prostate, salivary gland, seminal vesicles, skeletal muscle, skin, spinal cord, spleen, testes with epididymides, thymus, thyroid/parathyroid, trachea, urinary bladder, uterus, and gross lesions/tissue masses. In addition, gross changes/tissue masses were examined histologically from all animals in the lower-dose groups. Subsequently, the urinary bladder from males of the 1.25- and 2.5%-dose groups were also examined histologically.

During the premating period, no treatment-related effects were seen on mortality or body-weight gain but there was a dose-related increase in food consumption. After mating there were no treatment-related effects on the number of successful pregnancies or pup viability at birth, but pup mortality was increased in the 5%-dose group during the period 4-14 days of lactation. Mean pup weight was reduced in all treated groups, most markedly in the high-dose group.

In the F_1 generation, mortality was slightly higher in all treated groups than in contols, but it did not vary in a dose-related manner. Mean body weights of the high-dose males were consistently lower than controls, even though their food intake was elevated. Fasting blood glucose levels were elevated in females in all treated groups at 3 and 12 months, females in the 1.25 and 2.5% groups at 18 months, and males in all treated groups at 12 and 18 months.

At interim (12 months) sacrifice, the mean absolute and relative thyroid weights were elevated in the high-dose males while the relative kidney weights were elevated in the high-dose females. At termination, the thyroid weights were elevated in males of the 2.5- and 5%-dose groups and females of the 5% group; kidney weights were elevated in both sexes of the 5%-dose group and females of the 2.5% group. No treatment-related effects were seen in urinalyses, haematology determinations, physical observations, or ophthalmology.

Histopathological examination revealed an increased incidence of urothelial hyperplasia in treated males and of urinary bladder transitional cell/urothelial neoplasms in males of the 5%-dose group. The overall incidences in males are summarized below:

Group	Control 1	Control 2	1.25%	2.5%	5%
Number examined	58	61	58	55	60
Number with neoplasia	1	2	1	2	5
Number with hyperplasia	1	4	7	10	3

Non-statistically-significant increases in testicular Leydig cell tumours and neoplastic nodules in the liver were also observed. When time-to-tumour analysis was performed on pathology incidence data, the increased incidence of bladder tumours was confirmed and the incidence of several other tumour types showed statistically-significant differences related to treatment, including neoplastic nodules in the liver (males and females), female mammary adenomas and pituitary adenomas, male parthyroid adenomas, male thyroid medullary carcinomas, female uterine leiomyosarcomas, and male testicular interstitial/Leydig cell tumours.

Of the non-neoplastic pathology, chronic nephropathy was a common finding in all groups but the severity was greater in females in the 5%-dose group. Other lesions did not appear to be related to treatment (Knezevich & Hogan, 1981).

Special studies on mutagenicity

Fast Green FCF was non-mutagenic in the Salmonella/microsome assay (Brown et al., 1978) and negative results were also obtained in bacterial DNA repair tests (Kada et al., 1972; Rosenkranz & Leifer, 1980). The colour was inactive in a gene conversion assay in diploid yeast (Sankaranarayanan & Murthy, 1979).

In one of 2 experiments, colour-induced cell transformation occurred in cultured Fisher rat embryo cells at a concentration of 1 µg/ml (Price et al., 1978) and chromosome damage was reported in an in vitro test using Chinese hamster ovary cells (Au & Hsu, 1979).

Special studies on reproduction

Rats

A 3-generation reproduction study was carried out on Fast Green FCF in Long-Evans rats at dose levels of 0, 10, 100, 300, or 1,000 mg/kg b.w./day. The first generation parents (10 males, 20 females) were given the appropriate dose of Fast Green FCF in the diet 2 weeks before the first mating, and dosing continued throughout the gestation, lactation, and post-weaning phases for three successive generations. The F_0 generation rats were mated twice, the F_{1a} litters being necropsied at weaning, and selected animals (10 males, 20 females) from the F_{1b} litters were used for breeding.

Following an 80-day growth period, animals from the F_{1b} generation were mated 3 times and the offspring of the F_{2a} and F_{2b} generations were treated identically to the F_{1a} and F_{1b} generations. Following the third mating, half of the pregnant dams were sacrificed on day 19 of gestation, the uterine contents were examined for total embryos/resorption sites, and the corpora lutea per ovary were recorded. The other half were allowed to deliver normally (F_{2c}) and were sacrificed at weaning.

The F_{2b} animals were mated once and allowed to raise their offspring to weaning when both parents and offspring were culled.

Gross necropsies were performed on all parent animals and on F_{1a}, F_{2a}, F_{2c}, and F_{3a} offspring at weaning. Selected tissues from 5 animals of each sex/dose from the F_{1b} parents and the F_{3a} generation at weaning were fixed at necropsy, and the following tissues examined histologically from the control and high-dose group: stomach, ileum, jejunum, colon, liver, spleen, heart, lungs, adrenals, kidneys, urinary bladder, thyroid, ovaries, and uterus or testes.

No effects attributable to treatment were observed with respect to food consumption, body weight, adult mortality, mating performance, pregnancy and fertility rates, gestation length, offspring survival, weights and sex, litter survival, resorption rates, or necropsy findings. There were no macroscopic or microscopic tissue abnormalities of either F_{1b}- or F_{3a}-generation animals considered to be attributable to treatment (Smith, 1973).

Acute toxicity

Species	Route	LD_{50} (mg/kg b.w.)	Reference
Rat	Oral	> 2,000	Lu & Lavallee, 1964
Dog	Oral	> 200 mg/dog	Radomski & Deichman, 1956

Short-term studies

Dogs

Four beagles/group, equally divided by sex, were fed Fast Green FCF at 0, 1.0, or 2.0% of the diet for 2 years. Histopathology attributable to the colour was limited to green blobs of pigment in the renal cortical tubular epithelial cytoplasm of a male dog at the high-dose level; a female dog at the high-dose level showed slight interstitial nephritis and slight bone marrow hyperplasia (Hansen et al., 1966).

Long-term studies

Mice

Groups of 50 male and 50 female C_3HeB/FeJ mice were fed diets containing 1.0 or 2.0% Fast Green FCF for 2 years and 100 mice of each sex served as controls. After 78 weeks, 56 controls, 27 animals in the 1.0%-treatment group, and 17 animals in the 2.0%-treatment group still survived. Microscopic examination revealed no lesions that were attributed to feeding of the colour (Hansen et al., 1966).

Rats

Groups of 50 weanling Osborne-Mendel rats, evenly divided by sex, were fed diets containing 0, 0.5, 1.0, 2.0, or 5.0% colour for 2 years. No effects on growth or mortality were observed. Microscopic examination revealed no lesions that were attributed to the feeding of the colour (Hansen et al., 1966).

The colour was fed at a dietary level of 4.0% to 5 male and 5 female rats for periods from 18 to 20 months. This procedure resulted in gross staining of the forestomach, glandular stomach, small intestine, and colon. Granular deposits were noted in the stomach. No tumours were observed (Willheim & Ivy, 1953).

Observations in man

No data available.

Comments

The production of local sarcomata at the site of s.c. injection in rats is not considered to constitute evidence of carcinogenicity by the oral route. The mouse oral carcinogenicity study was negative but in the rat study, an increased incidence of urothelial hyperplasia and/or neoplasia of the bladder was observed. The biological significance of observed differences in benign and malignant tumours at other sites is questionable since, in some cases, statistically-significant differences were observed between the 2 control groups and, apart from the bladder, complete histological examination was not performed on the low- and intermediate-dose groups.

Biochemical studies have shown that the colour is poorly absorbed and the 3-generation reproduction/teratogenicity study was uneventful.

In view of the equivocal results of the most recent carcinogenicity study in rats, the evaluation is based on the earlier study, pending complete histological examination of all groups of rats and biometric examination of the data.

EVALUATION
Level causing no toxicological effect
Mouse: 5% in the diet equal to 18,600 mg/kg b.w./day
        falling to 8,000 mg/kg b.w./day.
Rat: 5% in the diet equivalent to 2,500 mg/kg b.w./day.

Estimate of temporary acceptable daily intake for man
0-12.5 mg/kg b.w.

Further work or information
Required by 1986
Complete histological examination of all dose-groups in the long-term carcinogenicity feeding-study in the rat and biometric examination of the data.

REFERENCES

Au, W. & Hsu, T.C. (1979). Studies on clastogenic effects of biologic stains and dyes, Environmental Mutagenesis, 1, 27.

Brown, J.P., Roehm, G.W., & Brown, R.J. (1978). Mutagenicity testing of certified food colours and related azo, xanthene and triphenylmethane dyes with the Salmonella/microsome system, Mutation Res., 56, 249-271.

Gangolli, S.D., Grasso, P., & Golberg, L. (1967). Physical factors determining the early local tissue reactions produced by food colourings and other compounds injected subcutaneously. Fd. Cosmet. Toxicology., 5, 601-621.

Gangolli, S.D., Grasso, P., Golberg, L., & Hooson, J. (1972). Protein binding by food colourings in relation to the production of subcutaneous sarcoma. Fd. Cosmet. Toxicology., 10, 449-462.

Grasso, P. & Golberg, L. (1966). Subcutaneous sarcoma as an index of carcinogenic potency. Fd. Cosmet. Toxicology., 4, 297-320.

Hansen, W.H., Long, E.L., Davis, K.J., Nelson, A.A., & Fitzhugh, O.G. (1966). Chronic toxicity of three food colourings: guinea green B, light green SF yellowish, and fast green FCF in rats, dogs and mice, Fd. Cosmet. Toxicology., 4, 389-410.

Hess, S.M. & Fitzhugh, O.G. (1953). Metabolism of coal-tar dyes. I. Triphenylmethane dyes. Fed. Proc., 12, 330-331.

Hess, S.M. & Fitzhugh, O.G. (1954). Metabolism of coal-tar dyes. II. Bile studies. Fed. Proc., 13, 365.

Hess, S.M. & Fitzhugh, O.G. (1955). Absorption and excretion of certain triphenylmethane colours in rats and dogs, J. Pharmacol. Exp. Ther., 114, 38-42.

Hesselbach, M.L. & O'Gara, R.W. (1960). Fast green and light green induced tumours: induction, morphology and effect on host. J. Nat. Cancer Inst., 24, 769-793.

Hogan, G.K. & Knezevich, A.L. (1981). A long-term oral carcinogenicity study of FD&C Green No. 3 in mice. Unpublished report No. 77-1781 from Bio/dynamics Inc., East Millsone, NJ, USA. Submitted to WHO by Certified Color Manufacturers' Association.

Iga, T., Awazu, S., & Nogami, H. (1971). Pharmacokinetic study of biliary excretion. II. Comparison of excretion behaviour in triphenylmethane dyes. Chem. Pharm. Bull., 19, 273-281.

Kada, T., Tutikawa, K., & Sadaie, Y. (1972). In vitro and host-mediated 'rec-Assay' procedures for screening chemical mutagens; and phloxine, a mutagenic red dye detected. Mutation Res., 16, 165-174.

Knezevich, A.L. & Hogan, G.K. (1981). A long-term oral toxicity/ carcinogenicity study of FD&C Green No. 3 in rats. Unpublished report No. 77-1780 from Bio/dynamics Inc., East Millstone, NJ, USA. Submitted to WHO by Certified Color Manufacturers' Association.

Lu, F.C. & Lavallee, A. (1964). The acute toxicity of some synthetic colours used in drugs and food. Canad. Pharm. J., 97, 30.

Nelson, A.A. & Hagan, E.C. (1953). Production of fibrosarcomas in rats at site of subcutaneous injection of various food dyes. Fed. Proc., 12, 397-398.

Price, P.J., Suk, W.A., Freeman, A.E., Lane, W.T., Peters, R.L., Vernon, M.L., & Huebner, R.J. (1978). In vitro and in vivo indications of the carcinogenicity and toxicity of food dyes. Int. J. Cancer, 21, 361-367.

Radomski, J.L. & Deichman, W.B. (1956). Cathartic action and metabolism of certain coal tar food dyes. J. Pharmacol. Exp. Ther., 118, 322-327.

Rosenkranz, H.S. & Leifer, Z. (1980). In Chemical Mutagens: Principles and Methods for their Detection. Ed: de Serres, F.J., & Hollaender, A. Plenum Press: New York & London, Vol. 6, p. 109.

Sankaranarayanan, M. & Murthy, M.S.S. (1979). Testing of some permitted food colours for the induction of gene conversion in diploid yeast. Mutation Res., 67, 309-314.

Smith, J.M. (1973). A three generation reproduction study of FD&C Green No. 3 in rats. Unpublished report No. 71R-736 from Bio/dynamics Inc., East Millstone, NJ, USA. Submitted to WHO by Certified Color Manufacturers' Association.

Willheim, R. & Ivy, A.C. (1953). A preliminary study concerning the possibility of dietary carcinogenesis. Gastroenterology, 23, 1-19.

SWEETENING AGENTS

HYDROGENATED GLUCOSE SYRUPS (HGS)

EXPLANATION

Hydrogenated glucose syrups (HGS) are a mixture of polymers of glucose obtained from starch by hydrolysis which, upon hydrogenation, results in chemical reduction of the end-group glucose molecule to sorbitol. HGS consists primarily of maltitol and sorbitol, with lower portions of hydrogenated oligo- and polysaccharides.

Hydrogenated glucose syrups were evaluated for acceptable daily intake by the Committee at its twenty-fourth and twenty-seventh meetings (Annex 1, references 53 & 62). At the twenty-seventh meeting, the Committee decided, on the basis of acute and subacute tests and reproduction and metabolism studies, to allocate a temporary ADI of 0-25 mg/kg b.w. to hydrogenated glucose syrups containing 50-90% maltitol. The Committee at that time requested that the results of a lifetime feeding study should be made available.

BIOLOGICAL DATA

Biochemical aspects

Absorption, distribution, biotransformation, and excretion

In male Swiss mice given single oral doses of either ^{14}C-maltitol or ^{14}C-glucose, expired ^{14}CO$_2$, faecal ^{14}C, and blood ^{14}C-concentrations were determined from 5 to 240 minutes post-administration. Labelled ^{14}C from maltitol was absorbed into the blood from the gastrointestinal tract much more slowly, exhaled substantially less, and excreted via the faeces significantly more than ^{14}C from glucose. Glucose and sorbitol, but not maltitol, were detected in the blood of treated animals (Kamoi, 1975).

Five to 10 male or female Sprague-Dawley rats/group were given single oral doses, by gastric intubation, of 200, 400, or 600 mg sorbitol, maltitol, or HGS (containing 6-8% sorbitol, 50-55% maltitol, 20-25% hydrogenated tri- to hexasaccharide, and 15-20% hydrogenated polysaccharide of more than 6 units) and were killed 1.5, 3, or 7 hours post-administration. Only a small quantity of non-hydrolysed maltitol crossed the intestinal barrier, since its concentration in blood was very small. Urinary excretion of sorbitol, maltitol, or HGS accounted for about 1% of the ingested quantity over a 7-hour period. Administration of increasing doses of maltitol or HGS resulted in increased concentrations of sorbitol and decreased concentrations of maltitol in the gastrointestinal tract, which demonstrates that the α-1,4 glucose-sorbitol linkage of maltitol is hydrolysed and liberates glucose and sorbitol (Verwaerde & Dupas, 1982a).

Two series of 10 male Sprague-Dawley rats (5 treated and 5 control rats) were housed individually in metabolic cages. The treated rats drank a solution of HGS (18% W/V) for 10 days, and thereafter pure water until sacrifice. The control animals received water only. One series of rats was sacrificed on day 11; the other series was sacrificed on day 21. Sorbitol and maltitol were determined daily in urine and plasma, and the liver, kidneys, and spleen were removed from the animals at terminal sacrifice. After maltitol ingestion per day ranging from 2.97-4.27 g in males, the maximum quantity of maltitol excreted per day was 6.27 $\pm$ 4.50 mg (the maximum excretion coefficient was 0.14%); in females, the amount ingested ranged from 2.46-3.03 g/day, and the maximum quantity of maltitol excreted per day was 2.04 $\pm$ 1.2 mg (the maximum excretion coefficient was 0.09%). The ingestion of HGS resulted in very small quantities of maltitol in the urine. Twenty-four hours following the removal of HGS from the rats, the quantity of maltitol excreted in the urine was reduced to essentially nil. Maltitol was absent in the plasma, liver, kidneys, and spleen of treated animals. In addition, the ingestion of HGS did not affect the sorbitol content in the examined organs (Verwaerde & Dupas, 1982b).

Groups of 35 male Wistar rats (about 200 g each) received by gavage 1.2 ml of a 50%-aqueous solution of maltitol, xylitol, sorbitol, or glucose, and their blood glucose and residual sugar alcohols in the digestive tract were determined hourly for 6 consecutive hours. Animals given the sugar alcohols exhibited lower blood glucose values than animals receiving glucose itself. Maltitol disappeared from the digestive tract the most quickly of the sugars that were investigated (Takae et al., 1972).

Weanling Wistar rats placed on diets containing 13 or 26% maltitol for 9 weeks had reduced body-weight gains and increased intestinal weights as compared with controls. Enzymatic tests in dosed rats indicated that the α-glycosidic linkage of maltitol was not hydrolysed with pancreatic enzymes or enzymes of the intestinal mucosa. Maltitol dehydrogenase was not observed in liver-cell cytoplasm and prolonged maltitol administration did not induce hepatic sorbitol dehydrogenase (Inoue, 1970).

Maltitol was administered to rats either by gastric probe at a dosage of 2 g/kg or i.v. at a dosage of 1 g/kg. Blood glucose and insulin levels and the liver glycogen content were measured at 4.5 hours. Maltitol induced hyperglycemia similar to that observed by the administration of an equivalent amount of glucose or sucrose (Lederer et al., 1974).

Ten grams maltitol was given to fasted germ-free or normal rats by stomach tube. Urine and faeces were collected for 24 hours after intubation and analysed for sorbitol and maltitol. No glucose was detected in the urine or faeces of any of the animals, and there was no significant difference in the maltitol content of the urine. The sorbitol content of the urine was higher in germ-free than in normal animals, but no statistical differences were found. In normal animals there was a 96% utilization of maltitol, while in germ-free animals, utilization was 84%, indicating that microbial utilization is not the major factor in maltitol metabolism. Maltitol injected i.v. (0.25 g/animal) was excreted almost quantitatively (88%) in the urine, producing no significant rise in blood glucose. No sorbitol was detected in the blood. The small amount of sorbitol in the urine indicated slight utilization of maltitol in body tissues (Kearsley et al., 1982).

Intestinal perfusions were carried out with 6 rats (WAG/ Rij), 5 females and 1 male, monoassociated with aerobic bacilli that are unable to metabolize maltitol. A solution (0.18 ml/min) containing 85 mM maltitol and 0.1% (w/v) polyethylene glycol 4000 was infused in the duodenum. Maltitol and polyethylene glycol were estimated in the lower ileum during a period of 4 hours. The portion of maltitol absorbed over the distance of the whole small intestine was found to be 19 $\pm$ 4%. Therefore, under the experimental conditions used, 2.9 μmol maltitol/min. was hydrolysed (the ingested amount was 15.3 μmol/min) (Zunft et al., 1983).

By stomach tube 5 ml of an aqueous solution of 30% (w/v, 4.36 mmole) maltitol and 1% (w/v) polyethylene glycol were introduced into the stomachs of 8 gnotobiotic rats. For comparison, the same amount of maltose was administered to 3 rats. The animals were killed by decapitation 60-120 min. after application. Maltitol, maltose, and polyethylene glycol were found only above the ileocaecal valve. The maltitol:polyethylene glycol quotient decreased in going from the stomach to the small intestine. Only 31% of the ingested maltitol was found in the ileum after 2 hours (Zunft et al., 1983).

Rats (groups of 3 or 6) were given a single oral dose of 1 or 2 g maltitol. Very little of either maltitol or sorbitol appeared in the faeces, but appreciable amounts of sorbitol found in the urine indicate that the maltitol had been hydrolised. When HGS was administered, the pattern of excretion was similar, but the quantities of both maltitol and sorbitol in the urine were significantly higher (Lian-Loh et al., 1982).

Excretion of maltitol and sorbitol were compared in germ-free and normal rats given oral doses of 2 g maltitol. Significantly less of both substances was recovered in the faeces of normal rats, but urinary excretion was similar in both groups. Three rats were injected with maltitol (250 mg/animal). The dose was virtually cleared from the blood within 1 hour, and at most the equivalent of approximately 30 mg was recovered as sorbitol in the excreta after 24 hours. There was no evidence of a rise in blood glucose after injection of maltitol. Little or no metabolism of maltitol had occurred in the tissues because hydrolysis was catalyzed by gastric or intestinal enzymes (Lian-Loh et al., 1982).

Wistar rats were dosed orally with a 20%-maltitol solution (320 mg/kg, containing about 1.28 μCi maltitol-U-^{14}C per 100 g b.w.). A second group was dosed with maltitol-U-^{14}C and fasted for 24 hours. Urine, faeces, and $^{14}CO_2$ were collected. The rate of $^{14}CO_2$ excretion in respired air of the fed group was more rapid than that of the fasted group; of the CO_2 excreted within 24 hours (18.0%), 90% was excreted within 8 hours after dosing. The $^{14}CO_2$ excretion per unit time was high in the fed group between 2 and 4 hours after dosing. The fasted group showed a high excretion rate between 2 and 6 hours after dosing, and the decline thereafter was more moderate than that observed with the fed group. About 22% of the dosed maltitol was hydrolysed in the fed group, while that of the fasted group was about 33%. Excretion in the urine of ^{14}C-derivatives was more rapid in the fasted group than in the fed group. About 50% of the ^{14}C excreted in the urine was maltitol, while the figure for the fed group was about 34%. The results show that a portion of the administered maltitol was not metabolized, but was absorbed in the intestinal tract and quickly excreted in the urine (Oku et al., 1981).

Five Charles-River rats (both sexes represented) were dosed with radioactive maltitol solution by intubation (125-260 mg/kg, 49.7 μCi). After dosing, 1 animal was placed in a breath collection chamber; the other 4 were placed in individual metabolism cages. Exhaled carbon dioxide was collected over a period of 14 hours. Urine and faeces were collected either over a period of 48 hours (2 rats) or 72 hours (3 rats). The rapid (1-2 hours) and appreciable appearance of $^{14}CO_2$ (45.5% in 14 hours) in the breath indicates that a portion of the maltitol was hydrolised in the stomach and the resulting components, glucose and sorbitol, were absorbed and catabolically utilized.

The calculated caloric utilization was 76%. Radioactivity in the urine varied from 3.94 to 9.40% in 72 hours. The faeces contained 4.38–13.22% of the ingested radioactivity at 72 hours. Thirteen percent of the total faecal radioactivity was due to volatile fatty acids (Rennhard & Bianchine, 1975).

HGS or maltitol (400 mg) was administered by gastric intubation to groups of 10 male rats (5 fasted and 5 fed). Rats were sacrificed after 7 hours. The tested products and their metabolites were then examined in the digestive tract and in the urine. The results showed that the α-1,4 glucose-sorbitol linkage was hydrolysed in rats, whether fasted or not. Seven hours after the administration of either HGS or maltitol, small quantities of free maltitol were detected in the digestive tract of fasted and fed rats (2.5 (fasted) and 3.8 mg after HGS administration, 6.2 (fasted) and 4 mg after maltitol administration). Sorbitol liberated by digestive hydrolysis of maltitol or HGS was more rapidly absorbed by fed rats than by fasted rats. However, this does not seem to affect maltitol or sorbitol urinary excretion (Verwaerde & Dupas, 1984).

Two male beagle dogs were given maltitol-U-^{14}C (51.2 μCi) by stomach tube. Blood samples were collected until 32 hours after dosing. The peak radiolabel concentration in plasma was 2 hours after maltitol administration (304 and 263 μg/ml, expressed as maltitol equivalent in the 2 dogs). The radioactivity present in the urine after 48 hours was 7.8 and 3.8% of the administered dose in the 2 animals (Rennhard & Bianchine, 1975).

In vitro investigations on maltitol utilization by human intestinal flora indicate that some strains of Lactobacillae, Enterobacteriaceae klebsiella, Bacteroidaceae, and Catanabacterium utilize maltitol (Mitsuoka, 1982a; 1982b).

In an in vitro study of ^{14}C-U-maltitol in everted intestinal sacs, the highest transport of ^{14}C-maltitol was displayed in the jejunum, followed by the ileum and duodenum. Twenty-four hours after oral administration of ^{14}C-U-maltitol, 60% of the radioactivity was detected in the caecum, large intestine, and faeces. Five percent was excreted in the urine and 1.2% was expired as CO_2 within 24 hours. When ^{14}C-U-maltitol was injected i.v., over 35, 60, and 85% of the administered dose was excreted in the urine within 1, 3, and 24, hours, respectively (Oku et al., 1971).

In vitro digestion of HGS and its main components was investigated using enzymes from human and rat intestinal mucosa for periods of 0.5 to 24 hours. The amount of maltitol present in HGS governed the enzymatic hydrolysis of its components. Maltitol was hydrolysed slowly by the intestinal mucosa of man (3.56%) and rat (7%) when compared with maltose, of which 76% and 26%, respectively, was hydrolysed (Verwaerde, 1982).

High disaccharidase activity for sucrose, maltose, and maltitol was found in the jejunum of female Wistar rats. Less was observed in the ileum and duodenum. Disaccharidase activity for maltitol was extremely low compared with that for sucrose and maltose. Kinetic data indicate that both maltose and maltitol compete for the same intestinal disaccharidase (Yoshizawa et al., 1975).

Pieces of New Zealand rabbit small intestine were everted and incubated with 100 mM substrate (maltitol, sucrose, or glucose). After removal at different times (20, 40, or 60 min.) of incubation, the volume of serosal fluid and the dry mass of the gut pieces were determined. Maltitol was hydrolysed, and the hydrolysis products were absorbed by the everted sacs. No maltitol was detected in the serosal fluid. The serosal glucose concentration increased at a slower rate after incubation with maltitol than after incubation with glucose or sucrose. The rate of hydrolysis and absorption decreased with the longer time of incubation (Zunft et al., 1983).

An enzymatic maltase/glucoamylase system isolated from the intestinal mucosa of the rat was used in an hydrolysis study. After 40 hours of incubation with different substrates, the percentage of hydrolysis was 100% for maltose, 91% for maltitol, and 79-95% for HGS. The maltase/glucoamylase system hydrolysed maltose more quickly than maltitol (200 and 15 nmoles/minute, respectively) (Montreuil et al., 1983).

Toxicological studies
Special studies on mutagenicity

HGS was evaluated in a series of assays for assessing the mutagenic potential of chemicals.

HGS did not induce a significant increase in genetic mutations in the Schizosaccharomyces pombe strain in a host—mediated assay (mouse) at doses up to 1 g/kg (Mondino et al., 1979a).

HGS was not mutagenic against 4 strains of Salmonella typhimurium at concentrations of 0.01 to 1 ml/plate (0.2—20 mg), with or without activation by rat liver microsomes (Fouillet et al., 1978a; 1978b; Hofnung, 1978).

Eight male Sprague—Dawley rats (4 controls) received daily by gavage doses of HGS equivalent to 0, 2.5, 5, 10, 15, or 20% in the diet for 15 consecutive days. Urine from each rat was collected on day 15. Urine that had been concentrated and purified was assayed for mutagenic activity against Salmonella typhimurium strains TA98, TA100, TA1535, TA1537, and TA1538 using the Ames test with or without meta-bolic activation. No mutagenic effects were found (Farrow, 1983d).

A micronucleus test was conducted in which HGS was adminis-tered orally to adult male mice at doses of 10 or 50 ml/kg on 2 consecutive days. The animals were killed 6 hours after the second administration of HGS. Femoral bone marrow was taken, and 2000 polychromatic erythrocytes per animal were counted and scored for micronuclei. The administration of HGS did not significantly increase the mean percentage of polychromatic erythrocytes carrying micronuclei (Siou et al., 1981).

HGS at concentrations of 0.5 to 1000 μg/plate was not mutagenic in the Ames test with Salmonella typhimurium strains TA98, TA100, TA1535, TA1537, or TA1538 with and without metabolic activation (Mondino et al., 1979b).

HGS did not induce a significant increase in the incorporation of ^{3}H-thymidine into human heteroploid fibroblasts at concentrations up to 300 µg/ml (Mondino, 1980).

HGS was added in vitro to C3H/10T 1/2 (Clone 8) mouse fibroblast cells using concentrations from 10 to 1000 µg/ml with or without metabolic activation. There was a significant increase in the prevalence of morphologically transformed foci due to exposure to HGS (Farrow & Sernau, 1982; Farrow, 1982c).

The potential of HGS to induce forward mutations at the thymidine kinase locus in the L5178Y mouse lymphoma cell line was assessed at concentrations ranging from 27 to 1000 µg/ml in the presence or absence of metabolic activation. HGS induced a slight increase in mutation frequency, but the increase was not dose-related (Farrow, 1982b).

The ability of HGS to induce mutations in Chinese hamster ovary cells in vitro was investigated at concentrations ranging from 49 to 4900 µg/ml. No significant increase in the frequency of structural chromosomal aberrations was seen at any of the concentrations tested, either with or without metabolic activation (Farrow, 1982a).

Special study on reproduction
Rats

A multigeneration reproduction study in Sprague-Dawley rats was conducted, in which males received 5.4-8.5 g HGS per day and females received 5.3-13.5 g of the same compound per day in drinking water as an 18% (dry weight/volume) aqueous solution for 3 successive generations, each comprising 2 consecutive litters. Concurrent controls were given water only. The inital number of animals in each generation was 10 males and 10 females.

During the study, food consumption was lower for treated animals than for controls, but without irreversible effect on the growth rate except for females of the F_2 generation. Animals from most groups may not have had decreased growth rates due to the caloric compensation of HGS in the drinking water. No abnormal behaviour was detected during treatment.

Among the organ weights recorded, the kidneys of treated rats had lower weights and caeca had higher weights than concurrent contols. Haematology and gross pathology of parents were not remarkable. Fertility, length of gestation, litter size, number of live and stillborn pups, post-natal survival, and the lactation index were not affected by treatment. Sex ratios of treated litters exhibited an approximate 10% decrease in the number of female pups (Leroy & Dupas, 1983).

Special study on teratogenicity
Rats

HGS was administered by gavage to groups of 30 pregnant Sprague-Dawley rats from day 6 to day 15 of gestation, inclusive, at dosage levels of 0, 3000, 5000, or 7000 mg/kg/day. A control group received the vehicle, distilled water, at the same dosage volume through the same period. On day 20 of gestation, all rats were killed and the number of corpora lutea in each ovary and the number, position, and condition of implantations were recorded. Viable foetuses were weighed, sexed, and examined externally. The thoracic and abdominal cavities of the remainder were dissected, examined, and then processed for subsequent skeletal examination. No adverse effects were observed on pregnant females or on litter responses (litter size, foetal weight, or pre- and post-implantation losses). No visceral or skeletal abnormalities in the foetuses attributable to the treatment were observed (Dupas & Siou, 1985).

Acute toxicity

Species	Route	LD_{50} (mg/kg b.w.)	Reference
Mouse, male & female	oral	24,370	Dupas, 1982a
Mouse	oral	16,000	Yamasaki et al., 1973a
Mouse, male	i.p.	10,640	Dupas, 1982a
Mouse, female	i.p.	12,430	Dupas, 1982a
Mouse	i.p.	18,500	Yamasaki et al., 1973b
Mouse, male	i.v.	6,390	Dupas, 1982a
Mouse, female	i.v.	8,150	Dupas, 1982a
Mouse	i.v.	12,000	Yamasaki et al., 1973b
Mouse	s.c.	24,000	Yamasaki et al., 1973b
Rat, male & female	oral	24,370	Dupas, 1982b
Rat, male & female	oral	24,000	Nishibori, 1968
Rat	oral	24,130	Kotani & Chiba, 1968
Rat, male & female	i.p.	13,000	Dupas, 1982b

Short-term studies

Rats

Groups of 20 Sprague-Dawley rats, evenly divided by sex, were fed diets containing 1, 15, or 20% HGS for 3 consecutive months. The diet of the controls was supplemented with 20% sucrose. No effects on mortality, growth, or food consumption were observed. No diarrhoea or other clinical symptoms were noted. No treatment-related effects were reported after ophthalmological examination, and organ weights were normal. Slight decreases in haemoglobin levels and erythrocyte counts occurred in both sexes of the 15%- and 20%-groups at week 4 and week 13 of treatment. Blood urea concentrations increased slightly in all treated females at week 4. Moderate elevation of blood urea and glucose were noted in all treated rats at week 13. Slight elevation of blood phosphous occurred in both sexes fed 15 and 20% of the test material at week 13 of the test. Gross and histopathological changes observed were not remarkable (Coquet et al., 1980).

Forty weanling Sprague-Dawley rats, equally divided by sex, were fed 20% HGS for 90 consecutive days. The same number of controls were fed a diet supplemented with 20% sorbitol. No mortality or note-worthy clinical symptoms occurred. No effects on haematology, blood chemistry, urinalysis, or organ weights were noted. Gross and micro-scopic examination revealed no histopathological alterations in organs or tissues at termination (Stevens et al., 1980).

Dogs
Four male or 4 female beagle dogs received 4.95 g/kg HGS every day for 13 weeks. Clinical examinations were performed daily and all animals were autopsied. Food intake in the treated animals was slightly decreased throughout the study, without any effect on the animals' body-weight gain. Ophthalmologic examinations did not reveal any abnormalities caused by the treatment. The only significant effects that were observed involved the occurence of diarrhoea. Haematological and urinary analyses did not reveal any abnormalities linked with treatment. At autopsy, no lesions signifying an effect of treatment were observed. Organ weights were not modified (Virat, 1982).

Long-term studies
Rats
Weanling Wistar-derived male rats were divided into groups of 10 each and fed 0, 5, 10, 20, or 30% of either maltitol or sucrose, or 20% HGS, for 31 weeks. Body weights were reduced at week 4 in groups of 20 and 30% maltitol and 20% HGS, but were similar between the

treated and control groups at week 8. At termination, body weights and selected organ weights of groups fed 5 to 30% maltitol were similar to rats fed 5 to 30% sucrose (Wada, 1972).

Groups of 15 male or 15 female young Wistar rats were fed diets containing 0, 1, 3, or 10% HGS for up to 13 months. Mortality, food and water consumption, body weights, haematology, blood chemistry, urinalysis, and organ weights were similar between the treated and control rats. At the 6-month interim sacrifice, about half of the treated rats exhibited dilation of the stomach. Slight edema of the colonic mucosa in half of the treated rats was observed at month 3 and in about 10% of the treated rats at month 6. At termination, the histopathological changes were comparable between the dosed and control animals (Yamasaki et al., 1973c).

Three groups of 26 male or 26 female Wistar rats were fed diets containing 0, 3, or 10% HGS for 78 weeks. An interim sacrifice of 3 males and 4 females per group was carried out at week 52. After week 50, mortality increased in both groups of dosed males, but only in the high-dose female group. Males of the high-dose group weighed substantially less than controls after week 60. Haematology, blood chemistry, and organ weights at interim and terminal sacrifices were similar between the dosed and control rats. The incidence of non-neoplastic lesions was not remarkable. Increased incidences of neoplasms of adrenal glands occurred in the dosed females and of the thyroid gland in the dosed males. Tumours of the skin and mammary region were not compound-related. Neoplasms of internal organs and tissues other than the above endrocine glands were not observed (Shimpo, 1977).

HGS was administered in drinking water to groups of 100 Sprague-Dawley rats (50 males and 50 females) at concentrations of 0 or 18% (w/v) for 24 months. The HGS consumption measured during the study was: 13.9 g/kg/day and 21.5 g/kg/day for males and females, respectively.

Weight gains in both groups were essentially identical, except for during a short preliminary period of adaptation. Subsequently, the treated female body weights were nearly always slightly higher than of the control females. The treated male body weights, which were slightly lower during the first year, increased during the second year to slightly exceed those of the control group.

Diarrhoea, which appeared from the first week of the study, and which decreased from the fourth week in treated animals, is usual in animals receiving drinking water with high osmotic properties.

Adaptation to the diet was also the cause of considerable caecum hypertrophy.

Haematological and clinical chemical parameters did not reveal changes related to the test material. The only significant variation observed in treated males and females as compared with the control grups was a decrease in urea levels due to considerably higher water consumption by treated rats.

Urinalysis did not show treatment-related differences.

Histopathological examination did not show any tissue changes related to HGS ingestion.

The spontaneous mortality rate recorded during the study was lower for treated males than for control males (5 vs. 13, respectively). Among females, mortality was slightly higher in the treated group than in the control group (11 vs. 8, respectively).

Neoplasms of internal organs and tissues occurred with the same frequency in control and treated animals (Dupas et al., 1984).

Observations in man

Six fasted normal volunteers (3 men and 3 women) and 4 diabetics (1 man and 3 women) were given orally 30 g of maltitol or 30 g of sucrose, and at 0.5-, 1-, 1.5-, 2-, and 3-hour periods post-administration the glucose and maltitol concentrations in blood and urine were analysed. Maltitol administration induced lower blood glucose levels than the sucrose administration, both in normal and diabetic volunteers. Urinary excretion of maltitol was also low (Atsuji et al., 1982).

Three normal volunteers, 3 diabetics, and 1 subject infected with acute hepatitis, were administered orally 50 g of glucose, sorbitol, or maltitol in aqueous solution, and at 0-, 0.5-, 1-, 2-, and

3-hour intervals post-administration the blood concentrations of these sugars were measured. Maltitol concentrations peaked at 0.5 hour and decreased sharply thereafter. Glucose followed a similar pattern, but its concentration was significantly higher than maltitol, especially in diabetics (Nishikawa, 1982).

Twenty-seven children, 3 to 14 years of age, consumed 18 to 42 g of hard candies made either from sucrose or HGS within one hour. At this dosage, the children developed slight nausea and "hate to eat more" symptoms, whether the candies were made from sucrose or from HGS. Only 3 children, from both groups, developed mild flatulence (Leroy, 1982b).

Twenty, 30, 40, or 60 g HGS or sucrose were consumed in bolus form by groups of 6-10 volunteers in a double-blind study. Ingestion of a single dose of 60 g HGS resulted in 80% of the volunteers reporting either abdominal discomfort, watery diarrhoea, colic, or an increase in flatus production. At lower doses, incidences of these symptoms were marginal. In another experiment, 10 different volunteers consumed intermittently 30, 60, or 120 g HGS or 64 g sucrose daily for 2 days in divided doses. The highest dose produced abdominal symptoms in 50% of the subjects, whereas at the lowest dose, symptoms occurred in about 20% of the volunteers. In yet another experiment, groups of 10 volunteers took either 30 g HGS or 30 g sucrose daily for 21 days, while groups of 12 subjects consumed 15 g HGS, 15 g sorbitol, or 15 g sucrose daily for 28 days. In all subjects taking HGS for 21-28 days, no adverse symptomology other than mild flatulence in 5 subjects at the higher dose was observed. Haematological and biochemical indices (liver function, glucose, cholesterol, lipoproteins, plasma insulin, and triglycerides) were not remarkable (Abraham et al., 1981).

Nine diabetics (5 men and 4 women) each received a single oral dose of 50 g maltitol, 50 g glucose, or 7 consecutive daily doses of 50 g each of sucrose, powdered starch syrup, or maltitol. At 0, 1, 2, and 3 hours post-administration the blood levels of glucose, immunoreactive insulin (IRI), free fatty acids (FFA), and triglycerides (TG) were determined. Single doses of maltitol produced lower blood glucose and IRI levels but higher FFA and TG concentrations than glucose administration itself. The presentation of results from the multiple administration are unclear for the purpose of evaluation (Takeuchi & Yamashita, 1972).

In a similar study, blood glucose and IRI concentrations exhibited flatter curves when 10 normal volunteers and 5 diabetics received a single oral dose of 50 g HGS than when 50 g glucose was administered (Yamakubi, 1971).

In a preference test on the sweetening power and quality of sweetness of HGS compared with sucrose, 20 young women preferred sucrose over HGS for its sweetening power, but both sugars scored identically in terms of the quality of sweetness (Yamakubi, 1971).

Eleven diabetics (6 men and 5 women) and 9 healthy control volunteers (7 men and 2 women) were given single oral doses of 50 g maltose or 50 g maltitol. Lower increases in blood glucose and IRI were observed with maltitol administration than with the administration of maltose. An additional 10 diabetics and 10 healthy control volunteers were offered "Zenzai", a thick bean syrup sweetened either with maltitol or with sucrose. Lower increases in blood glucose levels were observed in subjects that consumed "Zenzai" sweetened with maltitol than with sucrose. In the preference test, no significant differences were recorded in the intensity and quality of sweetness of "Zenai" sweetened with either substance. No unfavourable side effects were observed (Mimura et al., 1972).

The influence on carbohydrate metabolism of HGS (containing 89% maltitol) has been studied in healthy humans by measuring blood glucose and insulin levels after various oral dosages of HGS. Subjects given oral doses of glucose were used as controls. Six volunteers were given on different occasions 50 g of glucose or HGS at dosages of 10, 25, or 50 g. Blood glucose levels after the administration of maltitol were much less than after the administration of glucose; the area under the blood glucose curve after the administration of 50 g HGS represented about 25% of the area under the curve after glucose administration. Similarly, insulin levels were significantly lower after HGS ingestion than after glucose ingestion. At a dose of 10 g HGS, practically no increase in blood glucose or insulin was observed (Secchi et al., 1982).

Groups of 16 to 36 volunteers per dose level ingested daily 12 to 60 g of maltitol for 3 consecutive days. Ingestion of 50 or 60 g of maltitol induced diarrhoea in 15 and 30% of the volunteers, respectively (Nikken Chemicals, 1982).

Chemical changes in the blood induced by maltitol were compared with those induced by glucose in both healthy people and patients with several disorders, including diabetes mellitus. Blood glucose levels of healthy subjects were determined after the administration of glucose (12.5, 25, or 50 g) or maltitol (50 g). Based on the glucose absorption curve, 38% of the maltitol that was orally administered was absorbed through the intestinal tract, but the absorption of maltitol was more delayed than that of glucose.

After administration of maltitol to the groups with diabetes and the other diseases with impaired glucose tolerance, blood glucose levels were higher than in healthy people. Elevation of blood glucose levels by maltitol was 25 to 50% of the elevation by glucose, and there was a correlation between the elevation of blood glucose by glucose and by maltitol.

The peak blood glucose level after administration of a mixture of 50 g glucose and 50 g maltitol was lower than after the administration of 50 g glucose alone. Diarrhoea was frequently observed in healthy people and in patients with a borderline glucose tolerance test pattern after oral administration of maltitol. However, the frequency of diarrhoea was very low in patients with diabetes or the other disease with impaired glucose tolerance patterns (Kamoi et al., 1975).

Ten men and 7 women volunteers consumed daily 20 hard candies containing about 80 g/day of sweet product, for 2 consecutive weeks in a blind trial. Each volunteer was offered for 1 week sucrose control candies and for the other week the candies made from HGS. A

majority of volunteers complained during both trial weeks of digestive
disorders, such as loss of appetite, occasional diarrhoea, cramps, and
a "bloated feeling". At the dosage level of 80 g/day of hard candies
made from HGS, the limit of tolerance was exceeded (Leroy, 1982a).

Five normal men, 23 to 56 years old, ingested in the early
morning on an empty stomach 0.5 g/kg/day maltitol for 30 consecutive
days. On days 1, 7, and 30 at 1, 2, and 3 hours after the daily
administration of maltitol, the concentrations of maltitol and glucose
in the blood were measured and serum levels of protein, cholesterol,
bilirubin, uric acid, urea nitrogen, SGOT, SGPT, LDH, sodium,
potassium, and calcium were determined. Diarrhoea was not observed.
In 3 volunteers, blood glucose levels increased by about 20% 1 hour
after maltitol administration. No other changes were observed (Itoya
et al., 1974).

One hundred seven men (including 11 diabetics) and 20 women
(including 2 diabetics) were offered daily 30 to 180 ml of a 50% HGS
solution in 2 equal daily doses for periods up to 4 months. They were
examined daily and subjected to extensive monthly blood chemistry analyses. Occasional diarrhoea or accelerated intestinal transit occurred
at higher doses, more frequently in women than in men. Both sexes tolerated 30 ml/day for up to 4 months without any clinical or digestive
manifestations (Tacquet & Devulder, 1978).

Fifteen subjects received single oral doses of 50 g glucose
or 40-50, 80, or 100 g HGS. At 0-, 0.5-, 1-, 1.5-, 2-, 2.5-, and
3-hour intervals post-administration, the blood concentrations of
glucose and insulin were determined. Maltitol excretion was measured
in urine 3 hours after administration. Blood glucose and insulin
peaked at 0.5 hour after administration of either glucose or HGS, but
the increases after administration of HGS were less than after glucose
administration. The data on maltitol excretion in urine were not
conclusive (Debry, 1983).

Five healthy women and 5 diabetics (3 males and 2 females)
were offered single oral doses of either 50 g glucose or 50 g HGS on 2

different occasions. Another group of 5 normal volunteers (1 male and 4 females) and 5 diabetics (3 males and 2 females) were given orally 25 g sucrose, 25 g sorbitol, or 33 g HGS on 3 different occasions. At 0-, 0.5-, 1-, 1.5-, 2-, and 3-hour periods post-administration, the concentrations of serum glucose and insulin were determined. Administration of glucose produced the highest serum glucose concentrations. Insulin levels in normal volunteers given HGS varied between insulin levels in those volunteers administered sucrose and sorbitol. Among diabetics, there were virtually no differences observed after the administration of the different sugars (Vessby, 1982).

Thirty-five subjects (10 females and 25 males) were divided into 3 groups and given 50, 85, or 125 g/day HGS. Subjects were required to ingest each solution in 6 equal doses, 1 every 2.5 hours, and were asked to record any intestinal discomfort, flatulence or diarrhoea. Of the 12 subjects that consumed 50 g/day, 2 had diarrhoea, flatulence, and abdominal pain. At 85 g/day, 3/12 subjects had diarrhoea, flatulence, and abdominal pain. At 125 g/day, 6/11 subjects had diarrhoea, flatulence, and abdominal pain.

No differences were noted between male and female subjects. The authors concluded that over 85 g HGS could be tolerated without undue problems to most subjects if taken over a period of one day. Intestinal discomfort, flatulence, and diarrhoea increased in severity with intake of HGS (Kearsley et al., 1982).

Sixteen subjects (8 females and 8 males) who had fasted overnight received 0.5 g/kg of one of 5 different substances (HGS, maltitol, glucose, glucose and sorbitol in the ratios found in HGS, or high-maltitol syrup). No glucose was detected in the urine of any of the subjects after ingestion of any of the test carbohydrates. There were no significant differences in polyol levels in the urine among subjects administered HGS, the glucose/sorbitol mixture, or high-maltitol syrup. Blood glucose and serum insulin profiles indicated no differences among subjects administered HGS, maltitol, and the glucose/sorbitol mixture. All these substances induced lower peak values for glucose and insulin than did glucose. The results indicate that HGS and high-maltitol syrup are metabolized to approximately the same extent as their basic components (Kearsley et al., 1982).

Six male subjects were placed on a strictly-controlled diet containing 40% carbohydrate and 60% protein for 10 days. No lipids were included except when part of a protein source. Total daily caloric intake was about 1800 calories. The carbohydrate in the diet consisted of HGS, high-maltitol syrup, 43% glucose syrup, or glucose (the standard). Total sample intake was about 160 g/day, in 6 equal doses. The results of blood glucose and serum insulin analyses indicate that prolonged consumption of high-maltitol syrup or HGS leads to some adaption to these compounds as judged by elevated blood glucose and insulin peak values at the end of the trial. Diarrhoea and flatulence gradually disappeared after 4-5 days of consuming these hydrogenated samples. No problems were encountered with either glucose or 43% glucose syrup, and urine and faeces analysis revealed no carbohydrate present after ingestion of these samples. Subjects ingesting HGS or high-maltitol syrup excreted up to 10 g sorbitol and 0.5 g maltitol in the urine and up to 8 g sorbitol and 11 g maltitol in the faeces over the trial period. Over 99% of the hydrogenated test materials were retained in the body and were therefore presumably utilized. This was substantiated by the steady body weights exhibited by all subjects throughout the trial (Kearsley et al., 1982).

Two healthy volunteers (aged 30 and 35 years) each received a test meal of 69.5 g maltitol on an empty stomach. Twenty minutes after oral administration, blood glucose concentrations had increased by 20 and 30 mg/dl, respectively. They remained at this level for 2 hours and started to normalize 3 hours after maltitol application. At that time, the 2 individuals suffered from diarrhoea (2 and 3.5 hours, respectively) (Zunft et al., 1983).

Daily amounts of 35 g maltitol were given with meals for a period of 10 days to 4 subjects aged 34-53 years. Subjective signs (flatulence, gripes, and nausea) and faecal parameters (amount, frequency, pH, and content of maltitol) were compared with data from a control period without maltitol application. During the 10-day test period, there were no significant alterations of frequency and amount of faeces or of the pH. No maltitol was detected in the excreta by thin-layer chromatography (Zunft et al., 1983).

Four healthy men (21-23 years of age) were given daily doses of 10 g maltitol containing 79.65 μCi ^{14}C-U-maltitol for a period of 7 days. Expired breath, blood, urine, and faeces were collected. An average of 17% of the total $^{14}CO_2$ recovered was exhaled within the first 2 hours, and 43% of the total $^{14}CO_2$ was exhaled within 4 hours. The total recovery of $^{14}CO_2$ suggests a caloric utilization of maltitol in man of approximately 90%. This value is substantiated by the low radioactivity levels found in the faeces (5%) and by the presence of appreciable quantities of radioactive metabolites in the blood and urine 7 days after administration (Rennhard & Bianchine, 1975).

Comments

Several metabolic studies have been performed in rats and in man with HGS containing 50-90% maltitol. HGS is metabolized to glucose and sorbitol by disaccharidases in the intestinal mucosa. Data on the velocity of hydrolysis in comparison to the natural substrate maltose show that the maltitol hydrolysis rate corresponds to 5-7% that of maltose; maltose inhibits the hydrolysis of maltitol. Maltitol is absorbed in trace amounts, with a maximum of 0.05% of the ingested dose excreted as maltitol in human urine. Sorbitol, a hydrolysis product of maltitol, is absorbed very slowly. The second cleavage product, glucose, is produced at a slow rate due to the slow hydrolysis of maltitol; it is partially metabolized, like sorbitol, by the bacteria of the lower intestine. Sorbitol inhibits the absorption of glucose.

After the administration of ^{14}C-maltitol, 13% and 5% of the administered radioactivity was recovered in rat and human faeces, respectively; 13% of the total radioactivity found in rat faeces was represented by volatile fatty acids. The total urinary excretion of the administered radioactivity (4-5%) was comparable for rats and humans. A further indication that ^{14}C-maltitol is hydrolysed in the stomach and that the resulting components, glucose and sorbitol, are absorbed and catabolically utilized, is given by the rapid (1-2 hours) and appreciable appearance of $^{14}CO_2$ in the breath of rats. In rats, 43% of the total $^{14}CO_2$ was exhaled within 4 hours. The recovery of radioactivity as $^{14}CO_2$ within 24 hours of administration of ^{14}C-maltitol was 49% and 38-59% of the administered dose for rats and humans, respectively.

Studies in animals and humans revealed that HGS or its major component maltitol produced significantly lower blood-glucose levels and more stable insulin levels than glucose or sucrose due to slow metabolism of maltitol.

HGS was examined in in vitro and in vivo genetic toxicity assays. The results from the in vitro assays, with and without metabolic activation, suggest that HGS does not induce a mutagenic, clastogenic, genotoxic, or neoplastic transformation response. No in vivo clastogenic effects were observed.

Acute and short-term animal studies indicate that HGS is not toxic after single or repeated oral administration of large doses. In rats, no evidence of toxic effects of prolonged feeding of up to 15-20% of the diet was observed. A 90-day study in dogs showed no evidence of adverse effects, except for diarrhoea, at a level of 4.95 g/kg b.w./ day. A multigeneration reproduction study in rats, in which HGS was administered in drinking water as an 18% aqueous solution, did not reveal any toxicologically-significant effects. Human tolerance studies conducted with HGS in healthy and diabetic subjects showed a laxative effect at intake levels of 30-50 g/day.

EVALUATION

Estimate of acceptable daily intake for man

ADI "not specified". The fact that high doses of HGS exert a laxative effect in man, which is a common feature of polyols, should be taken into account when considering appropriate levels of use of polyols, alone and in combination.

REFERENCES

Abraham, R.R., Davis, M., Yudkin, J., & Williams, R. (1981). Controlled clinical trial of a new non-calorigenic sweetening agent. J. Human Nutr., 35, 165-172.

Atsuji, H., Asano, S., & Hayashi, S. (1982). A study on the metabolism of maltitol, a disaccharide alcohol. Unpublished report from Keio University, Japan. Submitted to WHO by Anic S.p.A.

Coquet, B., Rondot, G., & Mary, M.C. (1980). Thirteen-week oral toxicity study in the rat with Lycasin® 80/55. Unpublished report from Institut Français de Recherches et Essais Biologiques, l'Arbresle, France. Submitted to WHO by Roquette Freres.

Debry, G. (1983). Study of glycemia and insulinemia variations after per os absorption of a single Lycasin® 80/55 dose in man. Unpublished report from University of Nancy, France. Submitted to WHO by Roquette Freres.

Dupas, H. (1982a). Acute toxicity of Lycasin® 80/55: Determination of the LD_{50} in mice. Unpublished report from Roquette Freres, Lestrem, France. Submitted to WHO by Roquette Freres.

Dupas, H. (1982b). Acute toxicity of Lycasin® 80/55: Determination of the LD_{50} in the rat. Unpublished report from Roquette Freres, Lestrem, France. Submitted to WHO by Roquette Freres.

Dupas, H., Leroy, P., & D'Alayer, C. (1982). 24 months safety study of Lycasin® 80/55 on rats. Unpublished report from Roquette Freres, Lestrem, France. Submitted to WHO by Roquette Freres.

Dupas, H. & Siou, G. (1985). Lycasin® 80/55 teratogenic potential study in rats. Unpublished report from Roquette Freres, Lestrem, France. Submitted to WHO by Roquette Freres.

Farrow, M.G. (1982a). In vitro chromosome aberrations in Chinese hamster ovary cells with Lycasin®. Unpublished report from Hazleton Laboratories America, Vienna, VA, USA. Submitted to WHO by Roquette Freres.

Farrow, M.G. (1982b). Mouse lymphoma forward mutation assay: Lycasin®. Unpublished report from Hazleton Laboratories America, Vienna, VA, USA. Submitted to WHO by Roquette Freres.

Farrow, M.G. (1982c).   Cell transformation assay–Lycasin® (Batch
        E–644–B). Unpublished report from Hazleton Laboratories
        America, Vienna, VA, USA. Submitted to WHO by Roquette
        Freres.

Farrow, M.G. (1982d). Body fluid analyses (rat urine): Lycasin®.
        Unpublished report from Hazleton Laboratories America,
        Vienna, VA, USA. Submitted to WHO by Roquette Freres.

Farrow, M.G. & Sernau, R.C. (1982). Cell transformation assay
        Lycasin® (Batch E–644–B) with metabolic activation.
        Unpublished report from Hazleton Laboratories America,
        Vienna, VA, USA. Submitted to WHO by Roquette Freres.

Fouillet, X., Gutty, D., & Wolleb, U. (1978a). Report of experiments
        on the mutagen potential of Lycasin® 80/55 Ref. 61540
        (Ames test). Unpublished report from Battelle Institute,
        Geneva, Switzerland. Submitted to WHO by Roquette Freres.

Fouillet, X., Gutty, D., & Wolleb, U. (1978b). Report of experiments
        on the mutagen potential of Lycasin® 80/55 Ref. 62157
        (Ames test). Unpublished report from Battelle Institute,
        Geneva, Switzerland. Submitted to WHO by Roquette Freres.

Hofnung, M. (1978).   Products examined; Lycasin® Syrup (80/55).
        Unpublished report from l'Institut Pasteur, Paris, France.
        Submitted to WHO by Roquette Freres.

Itoya, N., Moriuchi, S., & Hosoya, N. (1974).   Effect of oral admin-
        istration of maltitol on human serum composition. J. Jap.
        Soc. Food Nutr., 27, 77–81.

Inoue, Y. (1970).   Effects of maltitol administration on the develop-
        ment of rats. Unpublished report from Nikken Chemicals Co.,
        Omiya, Japan. Submitted to WHO by Anic S.p.A.

Kamoi, M. (1975).   Study on metabolism of maltitol.   Part 1.   Funda-
        mental experiment. J. Jap. Diab. Soc., 18, 243–249.

Kamoi, M., Shimizu, Y., Kawauchi, M., Fujii, Y., Kikuchi, Y., Mizukawa,
        S., Yoshioka, H., & Kibata, M. (1975). A study on the
        metabolism of maltitol (Clinical study). J. Jap. Diab.
        Soc., 18, 451–460.

Kearsley, M.W., Birch, G.G., & Lian-Loh, R.H.P. (1982). The metabolic
        fate of hydrogenated glucose syrups. Stärche, 34, 279–283.

Kotani, S., & Chiba, S. (1968). Report of acute toxicity test of maltitol. Unpublished report from Juntendo University, Japan. Submitted to WHO by Anic S.p.A.

Lederer, J., Delville, P. & Cevecoevr, E. (1974). Study of a new sugar replacement: Maltitol from the Belgium sugar factory. 93, 311-9.

Leroy, P. (1982a). Report on the tolerance test of Lycasin® 80/55 administered in the form of sweets to adult human volunteers. Unpublished report from Roquette Freres, Lestrem, France. Submitted to WHO by Roquette Freres.

Leroy, P. (1982b). Report on the tolerance test of Lycasin® 80/55 administered in the form of sweets to children. Unpublished report from Roquette Freres, Lestrem, France. Submitted to WHO by Roquette Freres.

Leroy, P. & Dupas, H. (1983). Lycasin® 80/55: Three generation reproduction toxicity studies. Volume I – Study of reproductive functions. Volume II – Study of reproductive periods. Unpublished report from Roquette Freres, Lestrem, France. Submitted to WHO by Roquette Freres.

Lian-Loy, R., Birch, G.G., & Coates, M.E. (1982). The metabolism of maltitol in the rat. Br. J. Nutr., 48, 477-481.

Mimura, C., Koga, T., Oshikawa, K., Kido, K., Sadanaga, T., Jinnouchi, T., & Kawaguchi, K. (1972). Maltitol tests with diabetics. Jap. J. Nutr., 30, 145-152.

Mitsuoka, T. (1982a). Utilization of maltitol by intestinal lactobacillus. Unpublished report from Institute of Physical and Scientific Research, Toyko, Japan. Submitted to WHO by Anic S.p.A.

Mitsuoka, T. (1982b). Utilization of maltitol by intestinal flora. Unpublished report from Institute of Physical and Scientific Research. Toyko, Japan. Submitted to WHO by Anic S.p.A.

Mondino, A., Fumero, S., & Berruto, B. (1979a). Study of the mutagenic activity in vivo of the compound Marvie® with Schizosaccharomyces pombe. Unpublished report from the Institute of Biomedical Research, Ivrea, Italy. Submitted to WHO by Anic S.p.A.

Mondino, A., Fumero, S., Peano, S., & Berruto, P. (1979b). Study of
    the mutagenic activity of Marvie® compound with Salmonella
    typhimurium. Unpublished report from the Institute of
    Biomedical Research, Ivrea, Italy. Submitted to WHO by Anic
    S.p.A.

Mondino, A. (1980). Test of DNA non-programmed synthesis stimulation
    by the product Marvie® in human cell (EVE) cultures.
    Unpublished report from the Institute of Biomedical
    Research, Ivrea, Italy. Submitted to WHO by Anic S.p.A.

Montreuil, J., Bouquelet, S., Dupas, H., Verwaerde, F., & Rosiers, C.
    (1983). Study of intestinal maltase/glucoamylase in the
    rat. Unpublished report from Roquette Freres, Lestrem,
    France. Submitted to WHO by Roquette Freres.

Nikken Chemicals (1982). Diarrhoea after Malbit® administration.
    Unpublished report from Nikken Chemicals Co., Tokyo, Japan.
    Submitted to WHO by Anic S.p.A.

Nishibori, K. (1968). Study on acute toxicity test of maltitol. Unpub-
    lished report from Notre Dame Seishin University, Okayama,
    Japan. Submitted to WHO by Anic S.p.A.

Nishikawa, K. (1982). Variations of blood and urine sugar values with
    oral administration of maltitol. Unpublished report from
    Nishikawa Hospital, Tadera, Japan. Submitted to WHO by Anic
    S.p.A.

Oku, T., Inoue, Y., & Hosoya, N. (1971). Absorption and excretion of
    maltitol-U-^{14}C in rat. J. Jap. Soc. Food Nutr., 24,
    399-404.

Oku, T., Him, S.H., & Hosoya, N. (1981). Effect of maltose and diet
    containing starch on maltitol hydrolysis in rat. J. Jap.
    Soc. Food Nutr., 34, 145-151.

Rennhard, H.H. & Bianchine, J.R. (1975). Metabolism and caloric utili-
    zation of orally administered maltitol-^{14}C in rat, dog and
    man. J. Agric. Food Chem., 24, 287-291.

Secchi, A., Pontiroli, A.E., & Pozza, G. (1982). Influence of maltitol
    (Malbit®) on glycemia and insulinemia after oral
    administration. Unpublished report from Clinica Medica,
    Università di Milano, Italy. Submitted to WHO by Anic S.p.A.

Shimpo, K. (1977). Long-term toxicity of "Malti" a reducing malt millet jelly in particular reference to tumorigenic activity in rats. J. Toxicol. Sci., 2, 1-28.

Siou, G., Conan, L., & Pilatte, Y. (1981). Research of the possible mutagenic activity of Lycasin® by the micronucleus test on mouse. Unpublished report from Laboratoire d'Histopathologie et de Cytopharmacologie, Versailles, France. Submitted to WHO by Roquette Freres.

Stevens, K.R., Gagliardi, J.J., & Weinberg, M.S. (1980). Comparative study of HSH 80/55 and sorbitol in rats when materials are administered in feed for 90 days. Unpublished report from Booz, Allen and Hamilton, Inc., Florham Park, NJ, USA. Submitted to WHO by Roquette Freres.

Tacquet, A. & Devulder, A.B. (1978). Study of the clinical and biologi- cal tolerance of Lycasin® 80/55 in man. Unpublished report from Hôpital Calmette, Lille, France. Submitted to WHO by Roquette Freres.

Takai, Y., Ito, T., & Iwao, H. (1972). A study on the variations of digestion and blood sugar values with the lapse of time after sugar administration. J. Jap. Soc. Food Nutr., 25, 162.

Takeuchi, I. & Yamashita, M. (1972). Clinical test results of a disac- charide alcohol (maltitol). Eiyo to Shokuryo (Nutrition and Food), 25, 170.

Verwaerde, F. (1982). Digestion in vitro by enzymes of the intestinal mucosa of rat and man of Lycasin® 80/55 and of its main fractions with a comparison with several di- and polysaccharides. Unpublished report from Roquette Freres, Lestrem, France. Submitted to WHO by Roquette Freres.

Verwaerde, F. & Dupas, H. (1982a). Study of the in vivo digestion of Lycasin® 80/55 in the rat. Unpublished report from Roquette Freres, Lestrem, France. Submitted to WHO by Roquette Freres.

Verwaerde, F. & Dupas, H. (1982b). A study of the urinary excretion and risks of accumulation of maltitol in certain organs of rats fed with Lycasin®. Unpublished report from Roquette Freres, Lestrem, France. Submitted to WHO by Roquette Freres.

Verwaerde, F. & Dupas, H. (1984). Study of the in vivo digestion of Lycasin® 80/55 in the rat. Volume III. A comparative study on fed and fasting rats. Unpublished report from Roquette Freres, Lestrem, France. Submitted to WHO by Roquette Freres.

Vessby, B. (1982). Studies of Lycasin® as a sweetener in diabetes mellitus. Unpublished report from University of Uppsala, Sweden. Submitted to WHO by Roquette Freres.

Virat, M. (1982). A 13-week toxicity study of per os administered product (Lycasin® 80/55) in dogs. Unpublished report from Institut Français de Recherches et Essais Biologiques. Submitted to WHO by Roquette Freres.

Wada, F. (1972). The nutritional efficiencies of Maltit® (maltitol) and Maltit® syrup (hydrogenated malt - conversion starch syrup). Proc. 24th Jap. Food Health Assoc., Oct. 12-13.

Yamakubi, T. (1971). Malbit®: a new sweetener and its applications for diabetics Proc 14th Cong. Jap Soc. Diabetics. April 3-4.

Yamasaki, M., Tanabe, K., & Kimishima, K. (1973a). Acute toxicity of Malbit®: Examination after various means of injection. J. Yonago Med. Assoc., 24, 34-37.

Yamasaki, M., Tanabe, K., & Kimishima, K. (1973b). Acute toxicity of Malbit® in mice. Investigation by various means of injection. J. Yonago Med. Assoc., 24, 1-10.

Yamasaki, M., Tanabe, K., Matsumoto, Y., Tamaki, H., Kimishima, K., & Okamoto, S. (1973c). Toxicological studies of Malbit® in rats. J. Yonago Med. Assoc., 24, 10-30.

Yoshizawa, S., Moriuchi, S., & Hosoya, N. (1975). The effects of maltitol on rat intestinal disaccharidases. J. Nutr. Sci. Vitaminol., 21, 31-37.

Zunft, H.J., Schulze, J., Gärtner, H., & Grütte, F.-K. (1983). Digestion of maltitol in man, rat and rabbit. Ann. Nutr. Metab., 27, 470-476.

ISOMALT

EXPLANATION

Isomalt is an equimolar mixture of α-D-glucopyranosido-1,6-sorbitol (GPS) (sometimes called α-D-glucopyranosido-1,6-glucitol) and α-D-glucopyranosido-1,6-mannitol (GPM). Complete hydrolysis of isomalt yields glucose (50%), sorbitol (25%), and mannitol (25%).

Isomalt was evaluated at the twenty-fifth meeting of the Committee under the name "isomaltitol" (Annex 1, reference 56). On the basis of the data available, the Committee allocated a temporary ADI of 0-25 mg/kg b.w. Further results from lifetime feeding studies and multigeneration reproduction studies were required by 1985.

Since the previous evaluation these data have become available and are summarized and discussed in the following monograph. The previously-published monograph has been expanded and is reproduced in its entirety below.

BIOLOGICAL DATA

Biochemical aspects

Absorption, distribution, and excretion

Isomalt-^{14}C was administered orally to rats at doses of 250, 1000, or 2500 mg/kg b.w. Absorption of ^{14}C-activity was dose-dependent, varying from 80% in the low-dose group to 45% in the high-dose group. $^{14}CO_2$ in expired air 2 days after administration ranged from 62% of the administered dose in the low-dose group to 33% in the high-dose group. Elimination in the faeces ranged from 18-54%

of the administered dose over a period of 48 hours. Approximately 5% of the administered ^{14}C-activity appeared in urine. The authors pointed out that the $^{14}CO_2$ that they measured could have originated in 2 ways: from the expired air due to metabolism of glucose, sorbitol, and mannitol after their absorption from the gut, or from intestinal gases due to microbial fermentation in the caecum. Consequently, $^{14}CO_2$ excretion could not be used as a direct indication of energy utilization (Patzschke et al., 1975a).

The fate of isomalt in the gastrointestinal tract of female rats that had been adapted to the compound was investigated by increasing its dietary concentration from 10% to 34.5% over a period of 3-4 weeks. After administration of 1.7 g isomalt in 5 g feed, the contents of the stomach, small intestine, caecum, and large intestine were examined at intervals up to 6 hours. From the content of GPS, GPM, sorbitol, mannitol, and sucrose found in these organs, the authors concluded that GPS and GPM were only partially hydrolysed by the carbohydrases in the small intestine, while a substantial proportion of these compounds reached the caecum where further hydrolysis of glycosidic bonds occurred. Fermentation of the liberated hexitols occurred in the caecum, which was enlarged, and only small amounts of GPS, GPM, and hexitols reached the large intestine (Grupp & Siebert, 1978).

Small quantities of isomalt were detected in kidneys and urine of rats after they were given large oral doses of isomalt. On this basis, the authors concluded that unhydrolysed isomalt is absorbed to a small extent (Musch et al., 1973).

When isomalt was fed to rats for several weeks it was observed that faecal excretion declined steadily, while the caecum enlarged. The authors concluded that this resulted from adaptation and metabolism by the gut microflora. Similarly, during a 17-day feeding period in which 6 female rats received 3.5 g isomalt daily, the faecal content fell from 25% of the dose at the beginning to 1% at the end (Musch et al., 1973; Grupp & Siebert, 1978).

Renal clearance studies were conducted in adult female rats (250 g b.w.) infused with 1.8 g isomalt, GPS, or GPM over a period of 3 hours. Maximum plasma concentrations of 25 mM were obtained. These compounds were readily cleared and urinary concentrations of up to 100 mg/ml were recorded, which compares with a maximum urinary concentration of 0.6 mg/ml in rats receiving 5 g isomalt per day orally. After the infusion of either isomalt or GPS, free sorbitol was not detected in blood or urine, and blood glucose concentrations were unchanged, demonstrating the metabolic inertness of these disaccharide alcohols. From the infusion and excretion rates and the plasma concentrations that were observed, the authors concluded that GPS is distributed in extracellular water, but does not reach the intracellular compartments (Grupp & Siebert, 1978).

Rat intestinal maltase was shown to be active against isomalt, GPS, and GPM, but the rates of hydrolysis were slow (Grupp & Siebert, 1978; Musch et al., 1973).

The ratio of the rates of hydrolysis of sucrose isomalt-ulose, and isomalt by rat intestinal α-glucosidases was 100:30:12. Similarly, sucrose was hydrolysed about 20 times faster than GPS or GPM by disaccharidases from the small intestine of the pig, and the relative rates of hydrolysis of maltose, sucrose, isomaltulose and isomalt by human intestinal α-glucosidases were 100:25:11:2 (Gau & Müller, 1976; Grupp & Siebert, 1978).

Using isocaloric diets, a 30-day maintenance study was conducted on rats weighing 146 g and a 34-day growth study was performed on rats initially weighing 94 g. In these experiments carbohydrate (starch or sucrose) was partially replaced by isomalt, giving final concentrations of 34.5% in test diets. In the maintenance study, the food energy intake was 21% higher than in sucrose-containing diets and in the growth experiment the test group received 38% more energy in the feed than did control rats given starch. The fall in energy utilization was between 53% and 75%, depending on the test protocol (Grupp & Siebert, 1978; Siebert & Grupp, 1978).

Isomalt was fed to groups of rats (10 males and 10 females) at levels of 0, 2.5, 5.0, or 10.0% of the diet; a group fed 10% sucrose was used as an additional control. The following observations were made during weeks 73-75 of the chronic toxicity/carcinogenicity study: faeces production, moisture content and pH of faeces, apparent digestiblity of fat and protein, determination of isomalt and its break-down products (maltitol and sorbitol) in the faeces, excretion of lactic acid and short-chain fatty acids, and composition of the gut flora.

From the results, it seems clear that isomalt is largely digested or degraded in the gastrointestinal tract of the rat, since neither isomalt nor its degradation products, such as sorbitol or mannitol, could be detected in the faeces. If it is assumed that degradation takes place mainly by bacterial flora in the large bowel, one may expect an increased production of short-chain fatty acids, which generally result from bacterial fermentation. The fact that in the present study faecal excretion of short-chain fatty acids was not increased and the pH of the faeces was not decreased indicates that any short-chain fatty acids formed were absorbed through the gut wall, thus providing energy to the body.

The feeding of isomalt resulted in a slight increase in the production of faecal dry matter, accompanied by an increased excretion of protein (nitrogen). The increased excretion of protein is probably a reflection of an increased number of intestinal micro-organisms resulting from the supply of an easily-fermentable substrate, as isomalt probably is (Sinkeldam, 1983).

In the preparation phase of a metabolic study in rats, all the animals were adapted to isomalt by feeding them a mixture of 95% basal diet and 5% isomalt, then increasing the isomalt portion by increments of 5% every 5 days until it ultimately amounted to 30% of the diet. After 30 days, 24 rats were divided into 3 groups. The control group was fed a casein starch diet. In the other 2 groups this basal diet was supplemented with either 30% sucrose or 30% isomalt. Food consumption and faecal and urinary excretion were quantitatively measured for a period of 10 days. The prostprandial course of serum insulin activity was determined in all groups up to 7 hours after feeding.

The apparent digestibility and metabolisability of the energy of isomalt were 91.3% and 90.1%, respectively. These values were 7% and 11%, respectively, lower than those of sucrose. Faecal excretion of nitrogen after feeding isomalt was twice that of the sucrose group because of increased microbial activity. Isomalt and its hexitols were excreted in the urine, which represents a loss of potential metabolisable energy. Altogether, the increased losses of metabolisable energy after feeding isomalt as compared with sucrose amounted to 20-30% in this experiment. In contrast to the control and sucrose groups, there was no prostprandial increase in the activity of serum insulin; the authors concluded that this is due to reduced and delayed hydrolysis and absorption of isomalt in the gastrointestinal tract (Kirchgessner et al., 1983).

The 2 isomalt components, GPM and GPS, were assayed for glucose bioavailability using ketotic rats. With conversion rates into glucose of 6 and 20%, respectively, for free mannitol and sorbitol, 39% for GPM, and 42% for GPS, the metabolic glucose pool of the rat does not receive the full carbohydrate complement of these compounds. Under these conditions, 36% of the GPM and 32% of the GPS provided bioavailable glucose.; 50% is the theoretical maximum.

Less-than-theoretical bioavailability of glucose from isomalt was ascribed by the authors to microbial attack in the hindgut. The authors concluded that the data on rats were valid for other species demonstrating carbohydrate fermentation in the caecum and/or colon. Differences between GPM and GPS are caused by a differential delay of glucose absorption in the small intestine, which is also observed weth sorbitol. These studies point toward the important role played by the mammal-microbial symbiosis in the large bowel (Ziesenitz, 1983).

Groups of 12 ileum re-entrant fistulated pigs and 12 normal pigs (Dutch Landrace x large white) were fed 80% basal ration plus a combination of 10% isomalt and 10% sucrose, 20% isomalt, or 20% sucrose. After exposure to isomalt for 5 and 8 days, no isomalt could be detected in the faeces. Although no detectable quantities of sucrose reached the terminal ileum, 67% of the intact isomalt, plus mannitol and sorbitol, reached the terminal ileum in the group fed 20% isomalt; 54% reached the terminal ileum in the group fed 10% isomalt plus 10% sucrose. This means that 33% and 46% of the ingested amount of isomalt, respectively, was hydrolysed and absorbed in the small intestine.

In pigs fed 10%, and especially 20%, isomalt, the flow of the chyme along the small intestine was considerably accelerated during the first 3-4 hours after feeding, and the amount of chyme appearing at the terminal ileum was greatly increased compared with the animals fed 20% sucrose. This accelerated and increased flow of the chyme along the small intestine was ascribed by the authors as most likely due to the osmotic properties of the non-absorbed isomalt and its constituents.

Faecal digestibility of proteins and "nitrogen-free extract" in the 10%, and especially the 20%, isomalt diets was depressed compared with the 20% sucrose diet. As a result, energy digestibility with the isomalt diets was also depressed. The authors concluded that the lowered faecal digestiblity of the isomalt diets is most likely due to the increased excretion of bacterial mass resulting from the more intensive fermentation in the large intestine in the isomalt-fed groups (van Weerden et al., 1984a).

Fistulated and normal pigs were fed 10% sucrose between meals, 5 or 10% isomalt between meals, or 10% isomalt with meals. The passage and absorption rate of these substances were determined at the terminal ileum (10 pigs per treatment) or over the whole distance of the digestive tract (4 pigs per treatment). Ten percent sucrose was completely digested and absorbed in the small intestine. In the 3 isomalt treatments, 61-64% of the ingested compound passed the terminal ileum in the form of intact isomalt plus free sorbitol, free mannitol, and free glucose. None of these sugars was excreted in the faeces, indicating that isomalt and its constituents passing the terminal ileum are completely broken down in the large intestine. No influence of isomalt on the consistency of the faeces was observed. An increased flow of ileum chyme occurred in the period of 1-4 hours after administration of isomalt, which is probably related to the osmotic activity of non-absorbed isomalt and its constituents. Consequently, the ileal digestibility of the proteins of the basal diet was slightly negatively affected. The faecal digestibility of energy-containing compounds in the isomalt treatments was significantly lower than in sucrose treatment. This can be explained by the increased excretion of bacterial mass resulting from the more intensive fermentation in the large intestine in the isomalt-fed groups. The results of the determination of metabolisable energy of the diets indicate that the metabolisable energy value of isomalt is lower than that of sucrose (van Weerden et al., 1984b).

The in vivo metabolism of isomalt in the large intestine was simulated in an in vitro fermentation study to investigate its degradation using chyme from pigs as basic substrate additionally inoculated with faeces. In the first week, the fermentation of isomalt (3.65%) by non-adapted microflora was investigated. In the second week, isomalt fermentation by adapted microflora taken from pigs fed a basic diet supplemented with isomalt was studied. In the third week, both flora were studied in fermentation experiments with a high concentration of isomalt (7.30%).

Isomalt was degraded to lactic acid, volatile fatty acids, and gases (CO_2, CH_4, and H_2). The energy loss through these gases is of minor importance compared with the energy content of isomalt itself. Depending upon the concentration of isomalt (and possibly the type of flora), different amounts of intermediates (volatile fatty acids and lactic acid) were formed. These intermediates can be reabsorbed in the large intestine. A concentration of isomalt in chyme of 3.65% (which corresponds to 20% in the feed) did not lead to higher levels of intermediates in comparison with controls with no isomalt, suggesting that no extra energy take-up by the body occurs through these intermediates. A considerable part of the energy content of isomalt entering the large intestine probably leaves the body in the form of biomass (Bol & Knol, 1982).

Toxicological studies
Special studies on carcinogenicity
Mice

The carcinogenicity of isomalt was examined in a 2-year oral study in Swiss mice, by feeding the material at dietary levels of 0 (control), 2.5, 5.0, or 10% to groups of 50 male and 50 female mice. An additional control group was fed 10% sucrose.

General condition, behaviour, and mortality rates were not unfavourably affected by the test substance in any of the groups. All surviving male mice were killed during week 94, because mortality exceeded 80% in both control groups. The female mice, which showed better survival, remained in the study until the scheduled termination date in week 104. Mean body weights of females in the mid- and high-dose groups were relatively low from day 112 onwards. Mean food intake figures were comparable in all groups. There were no changes in haematological findings that could be ascribed to the feeding of isomalt. The absolute and relative weights of the caecum (filled and empty) were increased in both sexes fed 10% isomalt. Gross and microscopic examination did not reveal pathological changes that could be ascribed to the feeding of the test substance. High mortality in males was probably caused by infectious urogenital disease.

From the results of this study, the authors concluded that the feeding of isomalt at dietary levels up to 10% to mice throughout the major part of their lifetimes failed to show carcinogenic properties or any other effects of obvious toxicological significance (Dreef-van der Meulen et al., 1983).

Rats

Isomalt was examined in an oral long-term toxicity/ carcinogenicity study by feeding groups of 50 male and 50 female rats diets containing isomalt at levels of 0 (control), 2.5, 5.0, or 10% for a lifetime period (males, 128 weeks; females, 130 weeks). A diet containing 10% sucrose was included in the study as an additonal control. The rats were derived from parents that had been fed the same diets before mating and during the gestation and lactation periods (in utero exposure). Weanling rats obtained from the second mating cycle

(F_{1b}-rats) described in the "Special study of reproduction" were used for the carcinogenicity study.

Mortality rates were not affected in any of the treated groups. During the last half-year of the study, there was a tendency towards lower mortality in the group fed 5% isomalt and also, though to a lesser extent, in the group fed 10% isomalt, both in males and females. The overall mortality at the end of the study was 68% for males and 63% for females.

Body weights were relatively low in the males fed 10% isomalt during the major part of the study. However, the differences with the controls were generally less than 5%. In females fed 5 or 10% isomalt, body weights were generally slightly lower than those of the controls. The greatest differences were observed after 2 years of feeding, but amounted to no more than 10%. Food intake and food efficiency figures did not show any obvious differences among the various groups. However, males fed 10% isomalt tended to consume more food and to have lower food efficiencies than controls. Water intake figures were generally slightly higher in treated groups than in controls, but there was no clear dose-related response. Haematological findings did not reveal any effects related to the feeding of isomalt or sucrose. At a number of stages, animals fed isomalt excreted greater volumes of more diluted urine than the controls. Urine analyses and microscopic examination of the sediment did not reveal any treatment-related effects. Several blood plasma parameters showed statisticaly-significant changes in animals fed isomalt, but none of these changes were considered to be of toxicological importance. The relative weights of the caecum, filled or empty, were increased in rats of both sexes fed 10% isomalt. Macroscopic examination at autopsy did not reveal any abnormalities that could be ascribed to the feeding of isomalt.

Non-neoplastic and hyperplastic histopathological changes occurred in the kidneys of animals fed isomalt. These changes consisted of decreased nephrocalcinosis in the intercortico-medullary layer of females, increased nephrocalcinosis in the pelvis of males and females, and increased pelvic urothelial hyperplasia in females.

No indications were found of carcinogenic properties of the test substance.

The authors concluded that the feeding of isomalt at levels up to 10% in the diet of rats that had been exposed to the test substance _in utero_ and then continuously during their lifetimes did not induce any effects of obvious toxicological importance (Sinkeldam & Dreef-van der Meulen, 1983).

Special study on mutagenicity

Isomalt was non-mutagenic in the Ames test at concentrations up to 12,500 μg/plate (Herbold, 1978).

Special study on reproduction

Rats

A multigeneration study was conducted in groups of 20 male or 20 female rats to examine the effects of isomalt on reproductive performance. Rats were fed diets containing 0 (control), 2.5, 5.0, or 10.0% isomalt over 3 successive generations. One group of rats fed a diet containing 10.0% sucrose served as an additional control group.

Two litters from each generation were reared. Organ-weight analyses and histopathological examinations were made on selected offspring of the final (F_{3b}) generation. There were no outstanding differences in body weight, food intake, or food efficiency among the groups of parent animals (F_0, F_{1b}, and F_{2b}). A tendency toward lower body weights occurred in the group fed 10% isomalt, but only in males of the F_0 and F_1 generations. This tendency was also noticeable in the food efficiency figures. The fertility of females, number of pups per litter, general condition, appearance, sex ratio at birth, birth weight, growth rate, and mortality of the pups were not adversely affected by the feeding of isomalt. The resorption quotient did not indicate any embryotoxic effects of isomalt.

F_{3b} rats fed 10% isomalt for 4 weeks after weaning showed caecal enlargement, which was attributed to poor digestibility of the isomalt. Gross and microscopic examinations did not reveal any treatment-related pathological changes. A slight increase in the relative weight of the kidneys in males fed 10% sucrose was regarded as not being of toxicological significance.

The authors concluded that the feeding of isomalt at levels up to 10% in the diet to rats over 3 successive generations did not affect fertility, reproduction, or health and survival of the progeny (Sinkeldam & Dreef-van der Meulen, 1982).

Special studies on teratogenicity

Rats

Isomalt was fed to female Wistar rats from day 0 to day 21 of pregnancy at dietary levels of 0 (control), 2.5, 5.0, or 10%. No abnormalities in the condition or behaviour of the animals fed isomalt were observed during the experiment. Body weights, food intake, autopsy findings, organ weights, and litter data were comparable between the control group and all groups fed isomalt. Visceral and skeletal examination of the foetuses did not indicate any effects that could be related to the feeding of isomalt.

The authors concluded that under the conditions of the study, isomalt did not induce any embryotoxic or teratogenic effects in rat foetuses (Koëter, 1982).

Isomalt was administered in various concentrations from days 0-20 of gestation to groups of 25 female rats of the BAY:FB30 line (derived from Long-Evans). The feed mixture consisted of 90% Altromin basic feed plus 0, 2.5, 5.0, or 10% isomalt (supplemented with corn starch to a total of 100% in each case). Two other groups were established; one received a 10% sucrose-feed mixture and the other received Altromin basic feed at a level of 80% of the amount consumed by the control group (restricted diet). The daily food consumption of those rats fed 5% or 10% isomalt was reduced significantly; these animals did not consume much more food than the animals that received the restricted diet. The number of foetuses with retarded development was elevated as a function of dosage in the Caesarean-section groups that had received 5% or 10% isomalt. The foetal weight and the number of resorptions of embryos were slightly elevated in the group receiving a restricted amount of feed.

An adequate number of gravid animals were allowed to deliver their young vaginally in the control group and in the groups receiving 10% sucrose or 10% isomalt. They were then allowed to raise their young without any treatment. Some of the pups were raised to sexual maturity, mated within the groups, and then examined for delayed damage in fertility or reproductive performance.

Both sucrose and corn starch (in the control group) were tolerated without any signs of damage to the mother animals or their young. In the isomalt group, there was a reduction in food consumption during gestation (treatment period) and a reduction in weight gain of the mothers during this period and during the phase of nurturing the young. Prenatal losses and perinatal and postnatal mortality were elevated among the pups. The surviving pups exhibited normal development, however, and no signs of delayed damage were evident in the F_1 mating.

The question of whether these harmful effects were induced prenatally or postnatally was investigated in an experiment in which nursing mothers were exchanged, so that the control mother animals nurtured newborn pups from mother animals that had been fed 10% isomalt during gestation, and vice versa. Mortality of the pups treated with isomalt before birth remained elevated even when nurtured by control rats, whereas the newborn of untreated mothers were nurtured by the previously-treated mothers without any signs of toxicity. Therefore, damage to the foetuses due to isomalt must have occured prenatally.

The immediate great reduction in food consumption in those groups treated with isomalt led the authors of the study to the assumption that the mothers were impaired by the acute dosage of isomalt. Therefore, forced adaptation was accomplished over a period of 14 days by first offering the animals feed that contained 5% isomalt, and then 10% isomalt, as the only food source. At the end of this adaptation period, weight gains and feed consumption of the animals in the isomalt group were essentially the same as those in the control group. These animals were then mated and treated further with 10% isomalt until the 20th day of gestation.

In contrast to the earlier studies in which the mothers were not adapted to the diet, postnatal mortality in the isomalt group was not elevated. Weight gains and physiological development of the pups were comparable to those of both the concurrent controls and historical controls of this strain of rats.

The author concluded that the embryotoxic effects observed in this strain of rats were not a primary effect of the test substance,

but instead were a secondary embryotoxic effect due to maternal intolerance to the acute doses of isomalt at the beginning of gestation.
When this maternal intolerance was avoided by adaptation of the animals
to isomalt mixed with the feed before gestation, no embryotoxic effects
were observed (Schlüter, 1984).

Rabbits

Isomalt was fed to female New Zealand white rabbits from day
0 up to day 29 of pregnancy at dietary levels of 0 (control), 2.5, 5.0,
or 10%. No abnormalities in condition or behaviour of the rabbits were
observed during the experiment. Maternal performance was comparable in
all groups; fertility indices ranged from 65.7 in the control group to
75.0 in the 5.0% group and the gestation indices ranged from 88.5 in
the 2.5% group to 100 in the 5.0% group. Body weights, food intake,
autopsy findings, organ weights, and litter data did not reveal any
consistent or significant group differences. Visceral and skeletal
examination of foetuses did not indicate any effects that could be
related to the feeding of isomalt.

Under the conditions of the study, isomalt at concentrations
of 2.5, 5.0, or 10% in the diet was non-toxic to pregnant New Zealand
White rabbits and did not induce any teratogenic or embryo/foetotoxic
effects in rabbits (Koëter, 1983).

Acute toxicity

Species	Route	LD_{50} (mg/kg b.w.)	Reference
Rat	i.v.	> 2,500	Musch et al., 1973
	i.p.	> 2,500	Musch et al., 1973

Short-term studies

Rats

Groups of 15 male and 15 female rats received isomalt in the
diet at concentrations of 0, 3.3, 10, or 30% for 3 months; a similar
group received 30% sucrose in the diet. Appearance, behaviour, growth,
and mortality were unaffected in the 3.3% group. Rats receiving 10%

isomalt showed mild diarrhoea in the first 2 weeks, which ceased as the study continued; rats given 30% isomalt had severe diarrhoea in the first 2 weeks, which then diminished in intensity. Body-weight gains were impaired in the top-dose group, most markedly in males. Haematological parameters were unaffected by treatment after 5 and 12 weeks. After 5 weeks, male rats in the top-dose group displayed elevated plasma bilirubin levels and and lowered concentrations of urea and glucose; females in this dose group had raised alkaline phosphatase and glucose levels, while urea and protein concentrations were lowered. After 12 weeks, male rats in the top-dose group and females in all treatment groups had elevated plasma bilirubin concentrations; in the females, bilirubin levels increased in a dose-dependent way.

In both sexes, blood urea concentrations were significantly depressed at the highest-dose level. Blood oxalate concentrations were significantly elevated in males of all dose groups, but were within the range considered normal; no treatment-related changes were observed in blood cholesterol, uric acid, creatine, SGOT, or SGPT levels. Urinalysis at 5 and 12 weeks revealed no differences between control and treated rats. Autopsy of all animals did not reveal any treatment-related gross pathology, and organ weights were normal for the thyroid, thymus, heart, lungs, liver, spleen, adrenals, and testes or ovaries (caecal weights were not recorded). Kidney weights were lowered in the 30%-isomalt groups of both sexes, which may have resulted from reduced nitrogen metabolism. Histopathological examination was carried out on the heart, lungs, liver, spleen, kidneys, pituitary, thyroid, adrenals, testes, epididymis, prostate, seminal vesicle, ovaries, uterus, salivary glands, pancreas, oesophagus, stomach, intestine, lymph nodes, thymus, bladder, brain, eyes, aorta, trachea, skeletal muscle, bone (femur), and bone marrow (sternum). No treatment-related effects were seen.

In this study, dietary concentrations of up to 10% isomalt were tolerated without obvious organic damage. The authors claimed that, if the transient diarrhoea is taken into account, 3.3% dietary isomalt was well-tolerated. However, due to the elevated plasma bilirubin concentrations seen in female rats at all treatment levels, it is difficult to establish a no-effect level (Bomhard et al., 1978).

Female Sprague-Dawley rats recived diets containing 10% isomalt for 14 days without adverse effects (Siebert, 1972).

In other dietary studies, 10% isomalt caused transient diarrhoea that disappeared after adaptation; the feeding of isomalt was associated with caecal enlargement (Musch et al., 1973).

After caecetomy, diarrhoea persisted 4 to 5 times longer than in intact rats (Grupp & Siebert, 1978).

Dogs

Groups of 4 male and 4 female beagle dogs, 40-51 weeks of age, received isomalt at dietary concentrations of 0, 5, 10, or 20% for 13 weeks. No differences between control and test groups were observed in general behaviour or appearance; food intake and body weights were normal. Diarrhoea was observed in animals receiving 20% isomalt and, occasionally, in the 10%-dose group; normal faeces were produced by animals given 5% isomalt. Measurement of body temperatures, pulse rates, and reflexes and ophthalmoscopic investigations after 4, 7, and 13 weeks of treatment showed no treatment-related changes; haematological and clinical chemical parameters were normal at these times. Plasma urea concentrations were lower in the treated animals, sometimes significantly so, but still within the range considered physiologically normal. Urinalysis did not show treatment-related differences. At autopsy, no compound-dependent abnormalities were observed and organ weights were unaffected (the gastrointestinal tract components were not weighed). Histopathological examination did not detect any tissue changes related to the test material. Concentrations of intestinal tissue α-glucosidases (maltase, sucrase, and glucoamylase) were unchanged by treatment.

The authors concluded that concentrations of up to 20% isomalt in the diet did not produce any toxic injury. Allowing for the occasional ill-formed faeces in the 10%-dose group, the no-effect level was conservatively placed at 5% of the diet, equal to 1.67 g/kg b.w./day for 13 weeks (Hoffmann et al., 1978).

Long-term studies

Rats

A 1-year feeding study with isomalt in rats, which was part of a long-term toxicity/carcinogenicity study, was conducted. Isomalt was fed to groups of 10 male or 10 female rats at levels of 0 (control), 2.5, 5.0, or 10% in the diet. An additional control group was fed sucrose at a level of 10% in the diet. The rats were derived from parents that had been fed the same diets prior to mating and during the gestation and lactation periods (in utero exposure). Observations were made of general appearance and growth, and food and water intakes were measured. After 52 weeks, all rats were killed and examined for gross pathology. Seven different organs were weighed. Tissue samples of a wide range of organs were examined microscopically.

Body weights of males fed isomalt or sucrose tended to be higher than those of males of the control group during the experimental period. This is probably the result of relatively low body weights of the control animals at the initiation of the study. In females, body weights of the various groups were similar. Food intake of males fed isomalt or sucrose was generally slightly higher than that of the corresponding controls. In females, food intake of the various groups was similar. Food-efficiency figures did not show any obvious differences among the groups. There were no consistent differences in water intake between the test groups and the controls. Gross and microscopic examination did not reveal pathological changes which could be ascribed to the ingestion of the test substance. The only treatment-related change consisted of an increase in the relative weights of the filled and empty caeca in males fed 10% isomalt.

The authors concluded that isomalt, fed at levels up to 10% in the diet to rats that had been exposed to the test substance in utero and then continuously during a 1-year period, did not induce any effects of obvious toxicological importance (Sinkeldam et al., 1981).

Dogs

Isomalt was administered orally to groups of 8 Beagle dogs (4 males and 4 females), at concentrations of 0, 2.5, 5.0, or 10% during a 1-year chronic toxicity study. As controls, 1 group of animals was given 10% maize starch in place of 10% isomalt and another group was given 10% sucrose.

Treatment with isomalt did not affect appearance or behaviour, nor did it have an effect on food or water intake; it did not affect body-weight gains. Likewise, there were no differences in body temperature or pulse rates and neurologic and ophthalmoscopic examination results showed no differences between control and treated animals.

The only consequence of treatment with isomalt was an increased occurrence of pappy to liquid faeces at all dose levels; this effect was most pronounced in the animals fed 10% isomalt.

No evidence of blood lesions, nor any influence on coagulation, was observed in treated animals. Clinical chemical examination, gross pathological and histopathological liver examination, and comparisons of organ weights showed no indication of hepatic deficiency. There was no indication of renal deficiency on the basis of clinical chemical blood or urine examination, gross pathological or histopathological examination of the kidneys, or comparisons of organ weights. Apart from the increased occurrence of pappy to liquid faeces during treatment with isomalt, which the authors concluded on the basis of the results is not of toxicological relevance, concentrations of isomalt up to 10% administered orally over a period of 12 months were tolerated by the dogs without harm (Hoffman et al., 1981).

Observations in man

Six volunteers were each given 15 g ^{14}C-isomalt orally. Approximately 10% of the administered radioactivity was excreted in the faeces of 5 individuals. One volunteer had abnormally rapid gastro-intestinal transit (due to beer drinking) and excreted 40% of the radioactivity in faeces. Approximately 5% of the radioactivity was excreted in the urine, principally in the first 24 hours. Serum levels of radioactivity reached a maximum (the equivalent of 130 µg isomalt/ml) within 1 hour. Small amounts of unhydrolysed isomalt were found in the urine, indicating that a minor proportion of the dose was absorbed unchanged (Patzschke et al., 1975b).

After oral doses of 100 g isomalt, an average of 0.1% of the dose was excreted in the urine within 24 hours in 19 studies; after the administration of 50 g isomalt, an average of 0.04% was voided in urine within 2 hours in 37 studies (Siebert et al., 1975).

Four female and 2 male volunteers, aged 20-56, took 3 x 20 g isomalt daily for 8 days in various foods. On average, less than 0.2% was excreted in the urine as disaccharides and less than 0.02% as hexitols. Maximum values were less than 1% and 0.1%, respectively. Less than 0.5% GPS and GPM together were found in faeces on any of the 8 days, and the daily mean amount of hexitol was never more than 0.07%. Excretion levels in faeces did not change significantly throughout the study (Siebert, 1977).

Three colostomy patients fitted with ileostomy bags were given 30 g isomalt in 250 ml herbal or fruit tea at breakfast after fasting for 12 hours. An average of 58.9% of the dose was found in the collection bag, indicating poor absorption of isomalt from the small intestine (Kronenberg et al., 1979).

Healthy volunteers were given either 50 g (43 subjects) or 100 g (7 subjects) isomalt on fasting stomachs. There were no significant increases in blood glucose levels within 2 hours after treatment (Siebert et al., 1975).

Six healthy volunteers with mean body weights of 80.5 kg were treated in a cross-over trial with sucrose, isomalt, or a placebo at a dose level of 1 g/kg b.w. The test material was administered, after an overnight fast, in 400 ml rose-hip tea, and a normal breakfast was eaten 30 minutes later. Blood glucose and insulin concentrations were determined 0, 0.5, 1, 2, 4, and 6 hours after treatment. Sucrose produced the expected increase in blood glucose and insulin levels within 30 minutes, returning to fasting levels within the observation period. In contrast, after treatment with isomalt, blood glucose levels were similar to the levels after treatment with the placebo throughout the study. Serum insulin levels were also similar up to 4 hours after treatment with either the placebo or isomalt, but with isomalt serum insulin levels increased to twice the fasting levels between 4 and 6 hours after treatment (Keup & Pütter, 1974).

Tests on 8 healthy female volunteers given 50 g isomalt in 400 ml water after fasting overnight revealed practically no change in blood glucose levels in the subsequent 3 hours (Mehnert et al., 1977).

The effects of 30 g isomalt and 30 g glucose were compared in a 73-year-old male diabetic using the glucose tolerance test. Glucose produced a maximal increase in blood glucose concentration of 77 mg/100 ml after 2 hours compared with an increase of 14 mg/ml after the ingestion of isomalt (Jahnke & Gierlich, 1978).

In similar studies, 24 adult onset diabetics were given 50 g isomalt or glucose in 250 ml rose-hip tea using a randomized, cross-over protocol. In groups receiving glucose, blood glucose concentrations increased by 123-141 mg/100 ml in 90 minutes and maximum insulin levels increased by 22.6 to 25.1 mE/1; in contrast, after dosing with isomalt, the increase in blood glucose was 10.5-12.7 mg/100 ml and insulin levels were similar to fasting concentrations (Jahnke & Gierlich, 1979).

A randomized, cross-over study was performed on 12 tablet-dependent diabetics given oral doses of 50 g isomalt or 50 g fructose in 400 ml water. The results confirmed that isomalt caused only a small increase in blood glucose levels and had little effect on insulin levels (Mehnert et al., 1979).

Ten adult volunteers were given isomalt at doses of 250, 350, or 500 mg/kg b.w. at intervals of 2 days between doses. The dose of 250 mg/kg b.w. was well-tolerated by all the subjects; 350 mg/kg b.w. produced flatulence in 8 of 10 volunteers, 2 of whom had diarrhoea. At 500 mg/kg b.w., only 2 subjects did not have diarrhoea. The maximum tolerated dose for a single administration in aqueous solution was 250 mg/kg b.w. (Pütter & Spengler, 1975).

Four groups of 10 children aged 4 to 12 years were given doses of 15, 30, 45, or 60 g isomalt, spread over one day, in the form of sweets. One child at the lowest dose exhibited diarrhoea, but it was not clear that this was connected with treatment since children consuming doses of 30 g and 45 g tolerated isomalt without side-effects. Four children in the highest-dose group did not consume the full regimen. Of the remaining 6 subjects, 4 developed diarrhoea (Spengler, 1978).

Administration of daily doses of 3 x 20 g isomalt to 6 subjects for 8 days produced flatulence in the first 4 days, but this symptom subsequently subsided, indicating an adaptive improvement in tolerance (Siebert, 1977).

In a single-dose comparative study, 36 children aged 4 to 14 years were tested for their tolerance to isomalt or sorbitol. The test materials were administered at oral doses of 10, 20, or 40 g with breakfast. Diarrhoea was observed in 25% of the subjects receiving 40 g of either isomalt or sobitol, but not at lower doses. No differences in tolerance were observed between the 2 compounds (Spengler, 1979a).

Multiple-dose comparisons were made between isomalt and sorbitol over a period of 14 days during which 2 groups of 10 adult volunteers received daily doses of 50 g of either isomalt or sorbitol in 3 equal portions in the morning, at mid-day, and in the evening. In the isomalt group, mild diarrhoea was reported in 1 case on the sixth day; in contrast, sorbitol produced mild diarrhoea in 7 cases, and on several successive days. In this respect, isomalt was tolerated significantly better than sorbitol, although the degree of flatulence reported was similar in both groups. The symptoms diminished during treatment (Spengler & Schmitz, 1979).

Two subjects received daily doses of 50 g isomalt orally for 14 days. Faeces were examined microbiologically twice weekly over a period of 5 weeks (2 weeks prior to dosing, 2 weeks during dosing, and 1 week following). Stools were of normal consistency throughout, and isomalt had no significant effect on faecal pH or microflora. Both volunteers experienced flatulence in the first week of isomalt treatment, which diminished or disappeared during the second week (Linzenmeier, 1978).

Eight metabolically-healthy women each received 1 50 g oral dose of isomalt, while 24 type II diabetics (2 groups of 12 each) received either 50 g isomalt and 50 g fructose, or 50 g isomalt and 50 g sucrose, as part of 2 controlled cross-over studies. The levels of blood glucose, serum insulin, free fatty acids, lactate, and pyruvate did not change in the next 3 hours in the healthy volunteers. In the other 2 groups there were no increases in blood sugar or serum insulin levels after consuming isomalt, in contrast with observations after sucrose or fructose administration. Gastrointestinal syptoms (meteorism, flatulence, and mild diarrhoea) occurred in 2-4 persons in each of the 3 groups (Bachmann et al., 1984).

In a randomized cross-over study, the effects of oral loading with 50 g isomalt on blood glucose, urinary sugar, and serum insulin levels were compared with oral administration of 50 g glucose in 24 diabetics of the "maturity-onset" type. Isomalt, in contrast with glucose, did not cause any changes in blood sugar levels nor an increase in serum insulin levels. Glycosuria was reduced significantly after isomalt ingestion compared with after glucose administration. Diarrhoea and flatulence were reported more frequently after isomalt ingestion (45.8%) than after the ingestion of glucose (12.5%) under the experimental conditions (Drost et al., 1980).

Isomalt was compared to sucrose in a prospective double-blind controlled cross-over study. The acute effects of the oral ingestion of 30 g loads of isomalt or sucrose on plasma glucose, insulin, free fatty acids, lactic acid, and carbohydrate and lipid oxidation were studied over a period of 6 hours by means of continuous indirect calorimetry in 10 healthy normal-weight subjects (21-30 years). Unlike sucrose, the ingestion of which was followed by significant changes in plasma glucose, insulin, and lactic acid during the first 60 minutes of the test, no significant changes in these

parameters were observed following the administration of isomalt. The
increase in carbohydrate oxidation occuring between 30 and 150 minutes
was significantly lower (P < 0.01) following isomalt ingestion than
after the ingestion of sucrose. Conversely, the decrease in lipid
oxidation was significantly less (P < 0.01) after isomalt ingestion
than after sucrose ingestion. In contrast to other sugar substitutes,
no increase in plasma lactic acid was observed after isomalt
administration (Thiebaud et al., 1984).

In a randomized cross-over study, 24 type II diabetics were
first given 50 g isomalt and later 50 g glucose or vice versa in the
morning before breakfast. After the administration of glucose there
were definite increases in blood glucose, serum insulin, and C-peptide
concentrations. After isomalt ingestion, the rise in blood glucose,
serum insulin, and C-peptide concentrations were significantly less.
Gastrointestinal effects such as diarrhoea and flatulence were noted
only after a single high dose of isomalt. The authors concluded that
isomalt appears to be suitable as a sugar substitute in a diabetic
diet, since in comparison with glucose there were no significant
changes in blood sugar levels and additional insulin was not released
(Petzoldt et al., 1982).

In a controlled single-blind comparison study, 2 groups of
13 or 14 healthy, male adults received 24 g isomalt or sucrose per day
for a week. The incidence of subjective symptoms reported by the test
subjects after the ingestion of isomalt or sucrose, respectively,
were: diarrhoea, 15% versus 21%; flatulence, 38% versus 36%; and
stomach-ache, 31% versus 21%. All the test subjects in the isomalt
group were free of symptoms on day 8 (Spengler & Schmitz, 1983).

A controlled double-blind cross-over study was conducted to
compare tolerance to isomalt with tolerance to sorbitol after single
oral doses of 3 different amounts (10, 20, or 40 g) in 36 metabolically-
healthy children between the ages of 6 and 14. Single doses of 20 g
isomalt or sorbitol were tolerated without diarrhoea, but about 10% of
the children reacted with mild diarrhoea after receiving 40 g doses.
Two cases of flatulence were observed after the ingestion of 10 g of

either isomalt or sorbitol. No differences between the 2 substances were found (Spengler, 1979b).

In a 6-week controlled, randomized, cross-over study, 60 insulin-dependent children received their regular diet or their regular diet plus isomalt (20 g). The median age was 13 years. Twenty-six cases were available for evaluation from each half of the median at the end of the study. Daily doses of 20 g isomalt did not have any influence on serum insulin levels, the frequency of hypoglycaemia, or the incidence of glucosuria in type I diabetic children in either age group. Consumption was accompanied, however, by a mild increase in flatulence and soft or runny stools (Dorchy & Ernould, 1983).

Twenty-four grams isomalt were administered daily for 3 months to 12 type II diabetics managed by diet alone. Various parameters in these patients were compared with those of a control group of 12 diabetic patients managed by diet alone who were not given isomalt. The treatment with isomalt produced no differences in blood sugar (fasting or postprandial), haemoglobin levels, or serum levels of cholesterol, triglycerides, or high-density lipids compared with the control group. The results of hepatic and renal function tests were within normal limits before and at the end of the study in both groups (Pometta & Trabichet, 1983).

In a double-blind cross-over study, 200 healthy adult volunteers of both sexes received 50 g chocolate containing either 20 g isomalt or 20 g sucrose at 8 o'clock in the morning after a standard breakfast at intervals of 1 week. After isomalt ingestion, 16 of the volunteers (8%) reacted with diarrhoea, whereas none experienced diarrhoea after sucrose ingestion. The incidence of diarrhoea was 10.8% in female volunteers versus only 4.5% in male volunteers. The subjects had been informed of the possible gastrointestinal symptoms in advance, so it is possible that the expectations of the female volunteers and also their subjective evaluation of the resulting symptoms were different than those of the male volunteers. A higher frequency of defecation and increased flatulence were reported after isomalt ingestion compared with the ingestion of sucrose (Spengler et al., 1983).

During a double-blind test, 12 volunteers ingested sorbitol, and 12 others isomalt, in 10, 20, or 40 g doses administered in 1-2 week intervals. The internal neurological and cardiovascular examinations, as well as haematology and blood chemistry analyses, revealed no modifications in initial data or values which could be related to the intake of either of the 2 sugar substitutes. The intestinal symptoms (meteorism, flatulence, and diarrhoea) increased as the administered dose-strength increased. After 10 g of either substance, light abdominal pains were registered; after 20 g, these symptoms became stronger, and flatulence and diarrhoea occurred; the 10 patients having indicated no symptoms were equally divided between the 2 test groups. From a clinical viewpoint, intestinal symptoms were significantly higher in those volunteers given 40 g isomalt or sorbitol, which represents the normal consumption level of a sugar substitute, than in the other groups. The differences in response to isomalt and sorbitol were not significant (Spengler et al., 1979).

Isomalt, placebo, sorbitol, or sucrose (20 g dissolved in 200 ml water in the case of the placebo or 2 tablets of Natreen® dissolved in the same amount of water) was administered to 24 type I diabetics at 6 a.m., 10 a.m., 2 p.m., and 6 p.m. Each test substance was administered 6 times at each of these 4 periods. Average and maximum serum insulin levels and average and maximum blood sugar levels were significantly higher after sucrose ingestion than after ingestion of the other test substances. There were no significant differences between sorbitol and isomalt. Side effects reported after isomalt ingestion included vomiting in one patient, but diarrhoea was not reported with any of the test substances (Irsigler et al., 1984).

Comments

Hydrolysis of isomalt yields glucose (50%), sorbitol (25%), and mannitol (25%). Hydrolysis by intestinal disaccharidases in the small intestine is incomplete. Further metabolism by the microbial flora of the large intestine results in complete disappearance of the sweetener from the faeces. After the administration of ^{14}C-isomalt, excretion of radiactivity in expired air ranged from 33 to 62%, and in the faeces from 18 to 54% over a period of 48 hours, depending upon the dose; approximately 5% of the administered radioactivity appeared in the urine.

Isomalt was non-mutagenic in the Ames test.

A multigeneration reproduction study conducted in rats at levels up to 10% in the diet did not affect fertility or reproduction, nor did it affect the health or survival of the progeny.

It is unlikely that embryotoxic effects observed in strain FB30 rats were due to isomalt; these effects were probably the result of maternal impairment caused by the elevated acute doses of isomalt at the beginning of gestation. These effects were avoided by adaptation of the animal to isomalt when it was mixed with the feed before gestation. No embryotoxic or teratogenic effects were observed in Wistar rats or in New Zealand White rabbits fed at the same dietary levels as the strain FB30 rats.

The feeding of isomalt at levels up to 10% in the diet of rats exposed in utero and then continuously during 1 year did not induce any toxic effects. Concentrations of isomalt up to 10% administered orally over a period of 12 months were tolerated by dogs without harm, apart from the increased occurrence of liquid faeces.

Carcinogenicity was not demonstrated in rats that had been exposed to isomalt at levels up to 10% in the diet in utero and then continuously during their lifetimes. No evidence of carcinogenic properties of isomalt were observed after feeding mice at dietary levels up to 10% throughout the major part of their lifetimes.

Lifetime feeding studies of high doses of isomalt resulted in caecal enlargement in mice and rats and renal pelvic nephrocalcinosis in rats, effects common to other polyols.

Laxative effects in man were noted at 20-30 g/day.

EVALUATION

Estimate of acceptable daily intake for man

ADI "not specified". The fact that high doses of isomalt exert a laxative effect in man, which is a common feature of polyols,

should be taken into account when considering appropriate levels of use
of polyols, alone and in combination.

REFERENCES

Bachmann, W., Haslbeck, M., Spengler, M., Schmitz, H., & Mehnert, H.
(1984). Investigations of the metabolic effects of acute
doses of Palatinit®. Akt. Ernähr., 9, 65-70.

Bol, J. & Knol, W. (1982). In vitro fermentation of Palatinit®.
Unpublished report No. A 82.302/220828 from Centraal
Instituut voor Voedingsonderzoek (CIVO/TNO), Zeist, The
Netherlands. Submitted to WHO by Bayer A.G.

Bomhard, E., Luckhaus, G., & Muller, L. (1978). Palatinit® (Bay i
3930) subchronic toxicology investigations in rats. Unpublished pharmaceutical report No. 8025 from Bayer A.G.,
Wuppertal, F.R.G. Submitted to WHO by Bayer A.G.

Dorchy, H. & Ernould, Ch. (1983). Tolerance study with Palatinit® in
Form of Sweets for Diabetic Children. Unpublished report
submitted to WHO by Bayer A.G.

Dreef-van der Meulen, H.C., Woutersen, R.A., & Bruyntjes, J.P. (1983).
Life-span oral carcinogenicity study with Palatinit® in
mice. Unpublished report No. V 83.184/200299 from Centraal
Instituut voor Voedingsonderzoek (CIVO/TNO), Zeist, The
Netherlands. Submitted to WHO by Bayer A.G.

Drost, H., Gierlich, P., Spengler, M., & Jahnke, K. (1980). Blood
glucose and serum insulin after oral doses of Palatinit®
in comparison with glucose in diabetics of the adult type.
Verh. dtsch. Ges. Inn. Med., 86, 978-981.

Gau, W. & Muller, L. (1976). Unpublished data submitted to WHO by
Bayer A.G.

Grupp, U. & Siebert, G. (1978). Metabolism of hydrogenated palatinose,
an equimolecular mixture of α-D-glucopyranosido-1,6-
sorbitol and α-D-glucopyranosido-1,6-mannitol. Res. Exp.
Med. (Berl.), 173, 261-278.

Herbold, B. (1978). Palatinit® (Bay i 3930) Salmonella/microsome-
test to investigate point mutagenicity. Unpublished
pharmaceutical report No. 7578 from Bayer A.G., Wuppertal,
F.R.G. Submitted to WHO by Bayer A.G.

Hoffmann, K., Luckhaus, G., & Müller, L. (1978). Bay i 3930/
Palatinit®, subchronic toxicity study in dogs with administration in the feed (13-week feeding study). Unpublished
pharmaceutical report No. 7987 from Bayer A.G., Wuppertal,
F.R.G. Submitted to WHO by Bayer A.G.

Hoffmann, K., Luckhaus, G., & Müller, L. (1981). Bay i 3939
(Palatinit®) chronic toxicity in dogs by oral administration (a 12-months' feeding study). Unpublished pharmaceutical report No. 10299 from Bayer A.G., Wuppertal, F.R.G.
Sub- mitted to WHO by Bayer A.G.

Irsigler, K., Regal, H., Kaspar, L., & Spengler, M. (1984). Effect of
oral doses of Palatinit® on insulin consumption in type I
diabetics. Akt. Ernähr., 9, 60-64.

Jahnke, K. & Gierlich, P. (1978). Pilot study on Palatinit®. Unpublished report. Submitted to WHO by Bayer A.G.

Jahnke, K. & Gierlich, P. (1979). Comparison of effects of acute oral
doses of Palatinit® and glucose on blood glucose, serum
insulin, and other metabolic parameters in diabetics.
Research plan and statistical analysis of the results by
Bayer. Ph.D. dissertation and biometrics of 31.5.1979.
Unpublished report submitted to WHO by Bayer A.G.

Keup, U. & Pütter, J. (1974). Determination of blood sugar and plasma
insulin in healthy patients having absorbed an oral dose of
Palatinit® or saccharose. Unpublished pharmaceutical
report No. 4781 from Bayer A.G., Wuppertal, F.R.G.
Submitted to WHO by Bayer A.G.

Kirchgessner, M., Zinner, P.M., & Roth, H.-P. (1983). Energy metabolism and insulin activity in rats fed Palatinit®.
Internat. J. Vit. Nutr. Res., 53, 86-93.

Koëter, H.B.W.M. (1982). Oral embryotoxicity/teratogenicity study with Palatinit® in rats. Unpublished report No. V 82.101/222428 from Centraal Instituut voor Voedingsonderzoek (CIVO/TNO), Zeist, The Netherlands. Submitted to WHO by Bayer A.G.

Koëter, H.B.W.M. (1983). Oral embryotoxicity/teratogenicity study with Palatinit® in New Zealand White rabbits. Unpublished report No. V 83.237/221235 from Centraal Instituut voor Voedingsonderzoek (CIVO/TNO), Zeist, The Netherlands. Submitted to WHO by Bayer A.G.

Kronenberg, H.-G., Spengler, M., & Strohmeyer, G. (1979). Absorption of Palatinit®, an equimolecular mixture of α-D-glucopyranosido-1,6-sorbitol (GPS) and α-D-glucopyranosido-1,6-mannitol (GPM) in the small intestine of colostomy patients. Unpublished report from Bayer A.G., Wuppertal, F.R.G. Submitted to WHO by Bayer A.G.

Linzenmeier, G. (1978). The influence of 14 day oral administration of Palatinit® on the stool flora of 2 health male volunteers. Unpublished report submitted to WHO by Bayer A.G.

Mehnert, H., Haslbeck, M., & Bachmann, W. (1977). High level administration of Palatinit® to healthy volunteers. Unpublished report submitted to WHO by Bayer A.G.

Mehnert, H., Haslbeck, M., & Bachmann, W. (1979). Effectiveness of Palatinit® as a sugar replacement for diabetics from a medical point of view in comparison to fructose. Research plan and statistical analysis of the results by Bayer. Ph.D. dissertation and biometrics of 28.5.1979. Unpublished report submitted to WHO by Bayer A.G.

Musch, V.K., Siebert, G., Schiweck, H., & Steinle, G. (1973). Physiological-nutritional studies on the utilization of isomaltitol in rats. Zeitschrift fur Ernährungswissenschaft Suppl., 15, 3-16.

Patzschke, K., Weber, H., & Wegner, L. (1975a). Bay i 3930-^{14}C: Reabsorption und elimination - Research conducted on rats. Unpublished pharmaceutical report No. 5636 from Bayer A.G., Wuppertal, F.R.G. Submitted to WHO by Bayer A.G.

Patzschke, K., Wegner, L., & Horster, F.A. (1975b). Bay i 3930-^{14}C: Reabsorption und elimination – Experiments conducted on humans. Unpublished pharmaceutical report No. 5635 from Bayer A.G., Wuppertal, F.R.G. Submitted to WHO by Bayer A.G.

Petzoldt, R., Lauer, P., Spengler, M., & Schöffling, K. (1982). Palatinit® in type II diabetics. Dtsch. Med. Wschr., 107, 1910-1913.

Pometta, D. & Trabichet, C. (1983). Report on utilization of Palatinit® in type II diabetics treated by diet alone. Unpublished study from Cantonal Hospital of the University of Geneva. Submitted to WHO by Bayer A.G.

Pütter, J. & Spengler, M. (1975). Tolerance to a single dose of Palatinit® (Bay i 3930). Unpublished pharmaceutical report No. 5475 from Bayer A.G., Wuppertal, F.R.G. Submitted to WHO by Bayer A.G.

Schlüter, G. (1984). Palatinit®: Embryotoxicity studies on rats. Unpublished pharmaceutical report No. 12,451 from Bayer A.G., Wuppertal, F.R.G. Submitted to WHO by Bayer A.G.

Siebert, G. (1972). Personal communication submitted to WHO by Suddentsche Zucher A.G.

Siebert, G. (1977). Study on the reabsorption of Palatinit® in 6 human volunteers. Unpublished pharmaceutical report from Bayer A.G., Wuppertal, F.R.G. Submitted to WHO by Bayer A.G.

Siebert, G. & Grupp, U. (1978). α-D-Glucopyranosido-1,6-sorbitol and α-D-Glucopyranosido-1,6-mannitol (Palatinit®). Health and Sugar Substitutes. Proc. ERGOB Conf. Geneva, 109-113, Karger, Basle.

Siebert, G., Grupp, U., & Heinkel, K. (1975). Studies on isomaltitol. Nutr. Metabol., 18 (Suppl.1), 191-196.

Sinkeldam, E.J. (1983). Effects of Palatinit® ingestion on the gut flora and the gut contents of rats. Unpublished report No. V 83.007/212651 from Centraal Instituut voor Voedingsonderzoek (CIVO/TNO), Zeist, The Netherlands. Submitted to WHO by Bayer A.G.

Sinkeldam, E.J. & Dreef-van der Meulen, H.C. (1982). Multigeneration study with Palatinit® in rats. Unpublished report No. V 82.244/292310 from Centraal Instituut voor Voedingsonderzoek (CIVO/TNO), Zeist, The Netherlands. Submitted to WHO by Bayer A.G.

Sinkeldam, E.J. & Dreef-van der Meulen, H.C. (1983). Life-span oral toxicity and carcinogenicity study with Palatinit® in rats. Unpublished report No. V 83.412/292021 from Centraal Instituut voor Voedingsonderzoek (CIVO/TNO), Zeist, The Netherlands. Submitted to WHO by Bayer A.G.

Sinkeldam, E.J., Woutersen, R.A., & Hoetmer, A. (1981). One-year feeding study with Palatinit® in rats. Unpublished report No. V 81.416/212021 from Centraal Instituut voor Voedingsonderzoek (CIVO/TNO), Zeist, The Netherlands. Submitted to WHO by Bayer A.G.

Spengler, M. (1978). Palatinit®, tolerance test carried out at the Children's Clinic, City Hospital, Wuppertal-Barmen. Unpublished report submitted to WHO by Bayer A.G.

Spengler, M. (1979a). Palatinit® tolerance study (sorbitol control) on 36 metabolically healthy patients at the Children's Clinic, Barmen. Unpublished report submitted to WHO by Bayer A.G.

Spengler, M. (1979b). Palatinit®, tolerance study (controls: sorbitol) on 36 metabolically healthy patients at the Barmen Pediatrics Clinic. Unpublished report submitted to WHO by Bayer A.G.

Spengler, M. & Schmitz, H. (1979). Comparison of Palatinit® tolerance with sorbitol tolerance after 14-day oral administration to healthy adult patients. Unpublished pharmaceutical report No. 8449 from Bayer A.G., Wuppertal, F.R.G. Submitted to WHO by Bayer A.G.

Spengler, M. & Schmitz, H. (1983). Comparison of the tolerance of healthy male adults towards Palatinit® (generic name: isomalt) and sucrose after oral administration for 8 days. Unpublished pharmaceutical report No. 11930 from Bayer A.G., Wuppertal, F.R.G. Submitted to WHO by Bayer A.G.

Spengler, M., Sommer, J., & Schmitz, H. (1979). Comparison between tolerance to oral Palatinit® and sorbitol taken once in increasing doses by healthy adults. Unpublished pharmaceutical report No. 8457 from Bayer A.G., Wuppertal, F.R.G. Submitted to WHO by Bayer A.G.

Spengler, M., Schmitz, H., & Sommer, J. (1983). Comparison of the tolerance of healthy adults towards Palatinit® and sucrose after a single oral dosage. Unpublished pharmaceutical report No. 11892 from Bayer A.G., Wuppertal, F.R.G. Submitted to WHO by Bayer A.G.

Thiebaud, D., Jacot, E., Schmitz, H., Spengler, M., & Felber, J.P. (1984). Comparative study of isomalt of sucrose by means of continuous indirect calorimetry. Metabolism, 33, 808-813.

van Weerden, E.J., Huisman, J., & van Leeuwen, P. (1984a). The digestion process of Palatinit® in the intestinal tract of the pig. Unpublished report No. 528 from Institut voor Landbouwkundig Onderzoek van Biochemische Producten (ILOB), Wageningen, The Netherlands. Submitted to WHO by Bayer A.G.

van Weerden, E.J., Huisman, J., & van Leeuwen, P. (1984b). Further studies on the digestive process of Palatinit® in the pig. Unpublished report No. 530 from Institut voor Landbouwkundig Onderzoek van Biochemische Producten (ILOB), Wageningen, The Netherlands. Submitted to WHO by Bayer A.G.

Ziesenitz, S.C. (1983). Bioavailability of Glucose from Palatinit®. Z. Ernährungswiss., 22, 185-194.

THAUMATIN

EXPLANATION

Thaumatin is a mixture of intensely sweet proteins (thaumatins) extracted with water from the arils of the fruit of the West African perennial plant Thaumatococcus daniellii. The thaumatins have a normal complement of amino acids, except that histidine is not present. The molecular weights of the thaumatins are approximately 22,000 and their iso-electric points are in the range of 11.5-12.5. There are no unusual side-chains, atypical peptide linkages, or end-groups. Extensive disulfide cross-linking confers to thaumatin thermal stability, resistance to denaturation, and maintenance of the tertiary structure of the polypeptide chain. The maintenance of tertiary structure is critical to thaumatin's technical function. Cleavage of just one disulfide bridge results in a loss of sweet taste. (Iyengar et al., 1979).

Thaumatin is purified by selective ultrafiltration, but small amounts of organic non-protein impurities remain in the commercial product. These consist principally of the arabinogalactan and arabinoglucuronoxylan polysaccharides, both of which are normal constituents of plant gums and mucilages.

Thaumatin functions primarily as a flavour enhancer and as a high-intensity sweetener. The substance was previously evaluated by the Committee at its twenty-seventh meeting (Annex 1, reference 62). No ADI was allocated at that time, although specifications were prepared.

BIOLOGICAL DATA
Biochemical aspects
Digestion

In vitro studies with purified mammalian digestive enzymes, using a sequential enzyme system simulating the human digestive tract and a rapid multienzyme system, resulted in thaumatin being digested more rapidly than egg albumin (Hsu et al., 1977; Higginbotham, 1978).

A cross-over nitrogen balance study was performed over two 10-day periods using 2 groups of 10 male rats. During the first period, Group 1 was fed a semi-synthetic diet containing 10% protein from egg albumin, while a similar diet, but containing 5% egg albumin and 5% thaumatin, was fed during the second period. Animals in Group 2 received the diets in the reverse order. This procedure was used to compensate for the lack of histidine in thaumatin, which is an essential amino acid in the rat. The nitrogen digestibility of both diets was approximately 90%, but no allowance was made for the loss of endogenous nitrogen. The author concluded that the digestibility of thaumatin is at least equal to that of egg albumin. However, the biological value of thaumatin was lower than that of egg albumin, which reflects the lack of histidine (Edwards, 1981).

Toxicological Studies
Special studies on mutagenicity

At doses up to 50 mg/plate, thaumatin was not mutagenic in the Ames test using Salmonella typhimurium strains TA98, TA100, TA1535, TA1537, or TA1538 or in Escherichia coli strain WP2, either with or without S-9 mix (Higginbotham, 1980; Higginbotham et al., 1983).

In a dominant-lethal assay, groups of 15 male CD1 mice were intubated with thaumatin at 200 or 2000 mg/kg/day for 5 days, with trimethyl phosphate at 100 mg/kg/day for 5 days (positive control), or with the vehicle, distilled water (negative control). Following completion of treatment, each male was paired with 3 untreated females for a maximum of 7 days each week, and this was repeated for 7 consecutive weeks. In the positive control group, fertilization was unaffected but, during the initial 2 weeks following treatment, a marked loss of

embryos and foetuses occurred and the normality and rate of development
of zygotes were adversely affected. On the other hand, mating
performance and fertility of males, fertilization of ova, development
and survival of zygotes, and pre- and post-implantation losses were
unaffected by treatment with thaumatin, as assessed in animals
sacrificed after 4, 11, or 18 days post coitum. The authors concluded
that, under the conditions of the study, thaumatin did not induce
dominant-lethal mutations in the gametes of male mice (Tesh et al.,
1977a).

Special study on teratogenicity

Thaumatin was administered by gavage to groups of 20
pregnant CD rats from day 6 to day 15 of gestation, inclusive, at
dosages of 0.2, 0.6, or 2.0 g/kg/day. A control group received the
vehicle, distilled water, at the same dosage volume of 10 ml/kg
throughout the same period. On day 21 of gestation, all rats were
killed and the number of corpora lutea in each ovary and the number,
position, and condition of implantations were recorded. Viable
foetuses were weighed, sexed, and examined externally. The thoracic
and abdominal cavities of the remainder were dissected, examined, and
then processed for subsequent skeletal examination. No adverse effects
were observed in the pregnant females nor on litter responses (litter
size, foetal weight, or pre- and post-implantation losses). No visceral or skeletal abnormalities in the foetuses attributable to the
treatment were observed (Tesh et al., 1977b).

Special studies on allergenicity

Studies on ileum preparations from groups of 5 guinea-pigs
that had been sensitized by i.m. injections of either 50 mg thaumatin
or 50 mg egg albumin (together with incomplete adjuvant, which is
preferred for evoking an IgE antibody response), revealed that the
minimum dose of thaumatin capable of eliciting a response was 250 ng.
This is comparable to the minimum dose of egg albumin needed to evoke
an anaphylactic response in the gut of control animals sensitized with
egg albumin. Tests on ileum preparations from guinea-pigs sensitized
with protein and complete (Freund's) adjuvant gave an essentially
similar result, although thaumatin was slightly more sensitizing, 50 ng

evoking a response, whereas no effect was elicited with a similar dose of egg albumin (Stanworth, 1977).

Studies in rats that had been sensitized by s.c. injections of either 10 µg thaumatin or 10 µg egg albumin (together with Freund's adjuvant) showed that the levels of anaphylactic antibody in sera, determined by passive subcutaneous anaphylaxis dilution-titration, were uniformly lower with thaumatin than with egg albumin (Stanworth, 1977).

In vitro studies on the ability of thaumatin to initiate histamine release non-immunologically showed that approximately 1 mM thaumatin was needed to cause a release of 50% of the total available histamine from a purified rat-mast cell preparation. Synacthen (ACTH β^{1-24} polypeptide), the standard histamine-releasing agent used in this study, released 50% of the total available histamine at a much lower concentration, 2 µM. This finding was confirmed in vivo in normal baboons; intradermal injection of thaumatin at a concentration of 1 mM immediately after an i.v. injection of Evan's blue dye gave only a weak "blueing" response, whereas Synacthen gave a measurable response at a concentration of 1 µM (Stanworth, 1977).

Acute toxicity

Species	Route	LD$_{50}$ (mg/kg b.w.)	Reference
Mouse	oral	> 20,000	Ben-Dyke, 1975
Rat	oral	> 20,000	Ben-Dyke & Joseph, 1976

Short-term studies
Rats

In a preliminary 90-day study, groups of 10 male and 10 female CD rats were fed thaumatin at dietary levels of 1.0, 4.0, or 8.0% w/w. A control group received a basal diet supplemented with 8.0% casein w/w to compensate for the high protein intake.

No deaths occurred in any group. The body-weight gains of male rats fed 4.0 or 8.0% thaumatin were lower (6% and 9%, respectively) than those of casein-fed controls. Body-weight gains of female rats were not affected by treatment. Food consumption of male and female rats receiving 4.0 or 8.0% thaumatin was 5-11% lower than that of casein-fed controls. A statistically-significant reduction (3.2%, $p < 0.05$) in haemoglobin content in the blood of female rats fed 8.0% thaumatin was observed, although the values were within the range normally found for animals of comparable age and weight. Otherwise, the cellular and chemical composition of blood and urine were unaffected by treatment.

Tests for hepatic (bromosulfonphthalein retention) and renal function (concentrating ability) during week 12 in rats fed 8.0% thaumatin revealed responses that were similar to those obtained in the casein-control group. Using the Ouchterlony diffusion technique, no evidence was obtained for the presence of antibodies to thaumatin in the serum of animals treated with thaumatin for 13 weeks. Apart from a statistically-significant increase ($p < 0.05$) in both the absolute (14%) and relative (17%) liver weights of female rats receiving 8.0% thaumatin, detailed macroscopic and microscopic examination of a wide range of tissues revealed no changes which could be attributed to treatment (Ben-Dyke et al., 1976).

In a second 90-day study, groups of 20 male and 20 female CD rats received thaumatin at dietary levels of 0, 0.3, 1.0, or 3.0% w/w. During the treatment period, body weights and food consumption were recorded weekly. Additionally, the liver, kidneys, heart, lungs, spleen, and brain from all animals fed 0.3 or 1.0% thaumatin were examined microscopically.

No visible signs of reaction to treatment were observed. No animals died as a result of treatment with thaumatin. Body-weight values for males receiving 3.0% thaumatin were 6% higher during the study period than those of the control group and females receiving 1.0% thaumatin showed a 6% reduction in body weight from 6 weeks onward. Food consumption of females receiving 3.0% thaumatin was 7% lower than that of the controls throughout the treatment period. No treatment-related effects were observed on water intake or on ophthalmoscopic examination. Blood and serum analyses showed significant increases of 8 and 13% in haematocrits at week 12 in male rats fed 1.0 or 3.0% thaumatin, respectively, and significant reductions of 10 and 12% in haematocrits in female rats in the 0.3 and 1.0% groups, respectively. A dose-related reduction in triglyceride levels of up to 60% was observed in females during weeks 4 and 12 of treatment. Urinalysis was unaffected by treatment.

No treatment-related changes were seen during a detailed macroscopic examination of all animals at necropsy. An 8% increase in the kidney weights of treated females was observed; when related to body weight, the increase was 13%. Thyroid weights were significantly higher in all male treated groups than in controls and significantly lower in all female treated-groups than in controls, both in absolute weight and when related to body weight. The authors did not consider the changes in thyroid weights to be of toxicological significance, since the mean absolute thyroid weights of the control males were abnormally low compared with historical control values; the mean values for treated males were within the historical control values cited by the author. On the other hand, the mean absolute thyroid weights of control females were within the expected range when compared to historical controls, giving an apparent treatment-related decrease in absolute and body-weight-related thyroid weights. No treatment-related changes were seen in the tissue sections taken from the control and 3.0% groups, which were examined microscopically. A subsequent microscopic examination of thyroid tissues from all animals in all groups revealed no treatment-related changes (Hiscox et al., 1981; Wood, 1984).

Confirmation of the lack of hypo- or hyperthyroid effects was obtained from an additional study in CD rats. Two groups, 10 males and 10 females in each, were fed 3% thaumatin or 3% egg albumin in the diet for 4 weeks. Blood samples taken from the animals after the treatment period were analysed for thyroxine (T_4) and triiodothyronine (T_3). No statistically-significant differences in thyroid hormone levels were observed between animals fed thaumatin and those fed egg albumin. The authors concluded that thaumatin had no effect on thyroid function in rats at a dietary level of 3% (Danks et al., 1984).

Dogs

Groups of 4 male and 4 female beagle dogs were fed thaumatin in the diet for a minimum of 90 days at levels of 0, 0.3, 1.0, or 3.0% w/w. During the treatment period, body weights were recorded weekly and food consumption daily. At the end of the treatment period, a detailed macroscopic examination was performed and weights of the adrenals, brain, heart, kidneys, liver, lungs, ovaries, pituitary, prostate, spleen, testes, thyroid, and uterus were recorded. Microscopic examinations were performed on these and a wide range of other tissues from all animals.

No deaths occurred and there were no overt signs of reaction to treatment. Males receiving thaumatin showed slightly increased body weights relative to controls. Food consumption and water intake were unaffected by treatment. Ophthalmoscopic examination did not reveal any changes that could be related to treatment. Haematological examination of blood samples during weeks 4 and 12 of treatment revealed slight decreases in haemoglobin concentrations, erythrocyte counts, and haematocrits in males fed 3.0% thaumatin. However, the values were within the range determined from historical controls. Biochemical examination of blood samples did not reveal any treatment-related effects and urinalysis was unaffected by treatment. Macroscopic pathology at termination did not reveal any treatment-related changes. An increase in absolute liver weight (20%) was observed in males fed 3.0% thaumatin. When related to body weight, organ weights showed no treatment-related variations. Microscopic examination revealed no changes that the authors considered to be related to the administration of thaumatin (Barker et al., 1981).

Long-term studies
No information available.

Observations in man
The intense sweetness of the fruit of Thaumatococcus daniellii was first described by a British surgeon in the Pharmaceutical Journal (Daniell, 1855).

The long history of human use of the fruit as a sweetener, now largely displaced by sugar in urban areas, and the absence of unusual or toxic effects following its ingestion, is attested by numerous affidavits from village elders in Ghana and the Ivory Coast (Higginbotham & Stephens, 1984).

Thaumatin was assessed for oral allergenicity in humans by giving 100 mg/day thaumatin or lactose in gelatin capsules to 4 women and 6 men for a period of 14 days using a double-blind cross-over design. The volunteers were randomly assigned to 2 groups of 5 each and given either the test substance or lactose. All volunteers were prick-tested for common allergens and with a solution of thaumatin before the study. Seven volunteers were tested with thaumatin a second time before the study commenced to determine if sensitization to thaumatin could result from prick-testing itself. No sensitization was observed. At the completion of the study a further prick test was performed to determine if sensitization had occurred following thaumatin ingestion. No sensitization was detected. Blood was taken from the volunteers at the commencement and again after the 28-day study period. Examination of the sera for antibodies to thaumatin by the passive subcutaneous anaphylaxis technique in baboons and rhesus monkeys showed no reactions when challenged s.c. or orally with thaumatin. Clinical assessment showed no treatment-related allergic effects (Eaton et al., 1981).

Thaumatin was assessed for oral sensitivity and irritation in humans. Chewing gum containing 150 ppm thaumatin was administered to 25 volunteers, who each chewed five 5.3-gram gum sticks per day, each stick for 15 minutes, over a period of 28 days. A similarly-constituted group of 25 volunteers received untreated gum. Allocation of volunteers to test and control groups was random and under a double-blind code. No weal or flare reactions were noted in any volunteers after prick-testing either before or after the treatment period, nor were any visible signs of irritant or allergic responses detected on the oral mucosa after chewing either treated or untreated gum. The authors concluded that, under these conditions, thaumatin did not cause irritation of the oral mucosa or any allergic responses (McLeod et al., 1981).

A clincal study was conducted to determine the effects of thaumatin on haematological and blood biochemistry parameters. Eighteen male and 12 female volunteers were randomly assigned to 2 groups. Each participant was given a weekly supply of capsules, each containing either 280 mg thaumatin or 210 mg egg albumin, and asked to ingest one capsule each morning at 9 o'clock. This procedure was followed for 13 consecutive weeks. The capsules were coded and their composition was known only to the physician and consultant pathologist. Blood was collected and analysed during the week preceding the start of the trial and subsequently after 4, 8, and 12 weeks. No treatment-related changes in either the chemical or cellular composition of the blood were observed in volunteers consuming thaumatin when compared to the control group. The cumulative intake of thaumatin by these subjects was 25 g, which is some 140 times the estimated maximum consumer intake over this period (Tompkins & Enticknap, 1984).

Prick-testing of laboratory personnel who had inhaled thaumatin intermittently for periods up to 7 years showed that about one-half (67/140) responded to common inhalant allergens. A positive response to thaumatin was observed in 13 subjects, all except one of whom were atopic or allergic (Higginbotham et al., 1983).

Comments

There is no evidence that thaumatin is treated differently than other proteins with respect to hydrolysis or digestion. No antibodies to thaumatin were detected in either rats or humans after prolonged oral administration of quantities of thaumatin that substantially exceed the anticipated human exposure, thus indicating that the intact protein is not absorbed, and confirming the digestiblity of thaumatin. The possibility that hormonally-active polypeptides are present in digests of thaumatin, and that these may be absorbed intact and retain their activity, is unlikely because endocrine disturbances were not observed in toxicological studies.

Thaumatin showed no mutagenic or teratogenic effects and no allergenic effects were noted.

Variations in thyroid weights in a 90-day rat study (increases in males and decreases in females) revealed no treatment-related histological abnormalities; hypo- or hyperthyroid effects were not observed in a follow-up study in which statistically-significant differences in thyroid hormone levels (T_3 and T_4) were not observed.

Slight changes in haemoglobin concentrations, red blood cell counts, and packed-cell volumes observed in rats and dogs fed up to 3.0% thaumatin were not observed in a 13-week clinical study in human volunteers ingesting levels of thaumatin on the order of 140 times higher than the anticipated maximum daily intake, which has been calculated to be 1-2 mg/person/day.

The lack of toxicity, combined with its ready digestion to normal food components, indicate that thaumatin's only dietary effect is to make an insignificant contribution to the normal protein intake.

EVALUATION
Estimate of acceptable daily intake for man
ADI "not specified".

REFERENCES

Barker, J.D., Hiscox, D.N., & Wood, C.M. (1981). Talin protein: 90-day toxicity study in the dog by dietary admixture. Unpublished report No. TAL/2/81 from Toxicol Laboratories, Ltd., Ledbury, England. Submitted to WHO by Tate & Lyle PLC.

Ben-Dyke, R. (1975). Talin: Acute oral toxicity in mice. Unpublished report No. 75/TYL2/058 from Life Science Research, Stock, England. Submitted to WHO by Tate & Lyle PLC.

Ben-Dyke, R. & Joseph, E.C. (1976). Talin: Acute oral toxicity in rats. Unpublished report No. 76/TYL5/131 from Life Science Research, Stock, England. Submitted to WHO by Tate & Lyle PLC.

Ben-Dyke, R., Ashby, R., & Newman, A.J. (1976). Talin: Toxicity in dietary administration to rats for thirteen weeks. Unpublished report No. 76/TYL4/188 from Life Science Research, Stock, England. Submitted to WHO by Tate & Lyle PLC.

Daniell, W.F. (1855). Katemfe, or the miraculous fruit of Soudan. Pharm. J., 14, 158.

Danks, A., Hooks, W., Ashby, R., & Whitney, J.C. (1984). Talin: Four-week dietary study in rats to investigate thyroid function. Unpublished report No. TYL/073/TAL from Life Science Research, Stock, England. Submitted to WHO by Tate & Lyle PLC.

Eaton, K.K., Daniel, J.W., Snodin, D.J., Higginbotham, J.D., Stanworth, D.R., & Al-Mosawie, T. (1981). Talin protein: Assessment in man for oral allergenicity on challenge testing. Unpublished report submitted to WHO by Tate & Lyle PLC.

Edwards, D.G. (1981). Talin (Thaumatin): Nitrogen digestibility in the rat. Unpublished report No. B.128 from RHM Research Ltd., High Wycombe, England. Submitted to WHO by Tate & Lyle PLC.

Higginbotham, J.D. (1978). The digestibility of Talin protein in vitro. Unpublished report from Tate & Lyle PLC, Reading, England. Submitted to WHO by Tate & Lyle PLC.

Higginbotham, J.D. (1980). Mutagenicity testing of Talin protein sweetener in vitro. Unpublished report from Tate & Lyle PLC, Reading, England. Submitted to WHO by Tate & Lyle PLC.

Higginbotham, J.D., Snodin, D.J., Eaton, K.K., & Daniel, J.W. (1983). Safety Evaluation of Thaumatin (Talin protein). Fd. Chem. Toxicol., 21, 815-823.

Higginbotham, J.D., & Stephens, J.P. (1984). Food uses of Thaumatococcus daniellii in West Africa. Unpublished report from Tate & Lyle PLC, Reading, England. Submitted to WHO by Tate & Lyle PLC.

Hiscox, D.N., Hill, R.E., & Wood, C.M. (1981). Talin protein: 90-day toxicity study in the rat by dietary admixture. Unpublished report No. TAL/1/81 from Toxicol Laboratories Ltd., Ledbury, England. Submitted to WHO by Tate & Lyle PLC.

Hsu, H.W., Vavak, D.L., Satterlee, L.D., & Miller, G.A. (1977). A multienzyme technique for estimating protein digestibiltiy. J. Food Sci., 42, 1269-1273.

Iyengar, R.B., Smits, P., van der Ouderaa, F., van der Wel, H., van Brouwershaven, J., Ravestein, P., Richters, G., & van Wassenaar, P. (1979). The complete amino-acid sequence of the sweet protein thaumatin. I. Eur. J. Biochem., 96, 193-204.

MacLeod, G.L., Eaton, K.K., Daniel, J.W., Snodin, D.J., Higginbotham, J.D., & Waite, D. (1981). Assessment of oral sensitisation and irritation when formulated in peppermint chewing gum. Unpublished report submitted to WHO by Tate & Lyle PLC.

Stanworth, D.R. (1977). Preliminary assessment of the potential allergenicity of the sweet protein, Talin. Unpublished research report from the University of Birmingham, England. Submitted to WHO by Tate & Lyle PLC.

Tesh, J.M., Davidson, E.J., & Willoughby, C.R. (1977a). Talin: Test for dominant lethality in the male mouse. Unpublished report No. 77/TYL11/096 from Life Science Research, Stock, England. Submitted to WHO by Tate & Lyle PLC.

Tesh, J.M., Earthy, M., Tesh, S.A., & Willoughby, C.R. (1977b). Talin: Effects of oral administration upon pregnancy in the rat. Unpublished report No. 77/TYL10/179 from Life Science Research, Stock, England. Submitted to WHO by Tate & Lyle PLC.

Tompkins, G.D. & Enticknap, J.B. (1984). A comparison of the effects on the chemical and cellular composition of blood following the administration of thaumatin and egg albumin to human subjects for 13 weeks. Unpublished report submitted to WHO by Tate & Lyle PLC.

Wood, C.M. (1984). 90-day toxicity study in the rat by dietary admixture; further microscopic examination of thyroids from rats in study TAL/1/81. Unpublished report from Toxicol Laboratories Ltd., Ledbury, England. Submitted to WHO by Tate & Lyle PLC.

THICKENING AGENT

TRAGACANTH GUM

The Joint FAO/WHO Expert Committee on Food Additives has reviewed this substance many times in the past (Annex 1, references 19, 32, 44, 53, & 62). Toxicological monographs were prepared twice previously (Annex 1, references 20 & 33).

Since the previous evaluation, additional data have become available and are summarized and discussed here. Material from the earlier monographs is incorporated into this evaluation.

BIOLOGICAL DATA

Biochemical aspects

In a comparative study of the hypocholesterolemic activity of various mucilaginous polysaccharides, tragacanth gum fed at a level of 3% along with 3% cholesterol in the diet of cockerels inhibited the development of hypercholesterolemia (Riccardi & Fahrenback, 1965).

Tragacanth gum administered i.p., s.c., or per os 24 hours before hexobarbital had no effect on the hexobarbital sleeping time of mice. The effect of phenobarbital and urethan pretreatment to induce a shortening of hexobarbital sleeping time was blocked by i.p. injection of tragacanth gum, thus suggesting the presence of a hepatic effect of tragacanth gum (Fujimoto, 1965).

The influence of a number of hydrocolloids on the transit time of digesta, stool weight, and colour of stools was investigated in rats. Groups of 23 male rats were fed for 2 weeks the basal diet mixed with 2 or 20% of a thickening agent added at the expense of the entire diet. Tragacanth gum accelerated the digesta passage. All hydro-colloids tested gave the stools a lighter colour and increased their size and water content (Gohl & Gohl, 1977).

Ten Bacteroides species and several strains of anaerobic bacteria found in the human colon were surveyed for their ability to ferment 21 different complex polysaccharides. Many of the Bacteroides strains and a strain of Bifidobacterium (B. longum) fermented tragacanth gum (Salyers et al., 1977a & 1977b).

Fermentation of 20 polysaccharides by species of the family Enterobacteriaceae were examined. Species of Klebsiella, Serratia, and Yersinia fermented tragacanth gum. As a food additive, tragacanth gum may lose some of its properties when exposed in various ways to enteric organisms (Ochuba & von Riesen, 1980).

Female rats were dosed twice-daily with aqueous solutions of tragacanth gum at doses of 20, 40, and 80 mg/kg b.w. over a period of 4 weeks. All doses caused uncoupling of oxidative phosphorylation in liver and heart mitochondria and partial inhibition of mixed function oxidases of liver endoplasmic reticulum. The uncoupling was reversible in the case of heart mitochondria while it was progressive in liver mitochondria. Tragacanth gum did not have any adverse effect on the hepatic mixed-function oxidases at the 2 lower doses, while a 20% inhibition developed after 30 doses with 2 x 40 mg/kg (Bachmann et al., 1978).

Male albino Wistar rats were fed diets containing 0, 0.5, 1.5, 2.5, and 3.5 (w/w) tragacanth gum for 91 days. Microsomal protein and PL-480 content of the liver were measured. No compound-related effects were observed. Electron microscopy of liver and heart muscle from the treated rats showed no abnormalities in any of the test animals (Anderson et al., 1984).

Toxicological studies

Special studies on mutagenicity

Tragacanth gum was evaluated for genetic activity in the following in vitro microbial assays, with and without activation: Salmonella typhimurium (strains TA1535, TA1537, TA1538, TA98, and TA100) and Saccharomyces cerevisiae strain D_4. No mutagenic activity was observed in any of these assays (Litton Bionetics, 1977).

Tragacanth gum was not mutagenic in a number of tests using mammalian systems. These included: (a) Host Mediated Assay in vivo in rats and mice using Salmonella typhimurium strain TA1530 and G46 or mitotic recombination frequency in S. cerevisiae D3, (b) a cytogenic study in vivo of rat bone-marrow cells, and (c) an in vitro study with human lung cells (wt. 38) in tissue culture (Litton Bionetics, 1972).

Special studies on teratogenicity

I.p. injection of 1 ml 1% aqueous mucilage of Persian traga-canth gum (single dose or 5 doses of 0.2 ml each) into mice between days 11 and 15 of gestation caused the death of all foetuses. Oral or s.c. administration had no effect. All samples were found to be contaminated with Enterobacter spp. and the embryotoxic effects were attributed to bacterial metabolites (Frohberg et al., 1969).

Tragacanth gum showed no evidence of maternal toxicity or teratogenicity after oral administration (as a suspension in corn oil) at levels up to 1200 mg/kg b.w./day to pregnant mice (days 6-16 of gestation) or to pregnant hamsters at dose levels up to 900 mg/kg b.w./day (days 6-10 of gestation). Similar studies with pregnant rats at dose levels up to 1200 mg/kg b.w. (days 6-15 of gestation) and with pregnant rabbits at dose levels up to 700 mg/kg b.w. (days 6-18 of gestation) resulted in significant maternal mortality in rats at the 1200 mg/kg b.w. dose level and in rabbits at dose levels of 150 and 700 mg/kg b.w. At autopsy, the gross pathological finding was marked haemorrhage in the mucosa of the small intestine. Offspring from animals surviving in the high-dose group as well as those in other test groups showed no compound-related abnormalities in the soft or skeletal tissues (FDRL, 1972).

A study was done using a chick embryo test system. Tragacanth gum dissolved in 0.12 N HCl was injected either into the air sac or the yolk of fertile chicken eggs at dose levels up to 7 mg/kg. The administration of tragacanth gum did not result in a significant increase in mortality. All hatched chicks appeared normal. Abnormalities observed in eggs that failed to hatch were 22% test, 14% solvent-control, and 3.41% flock background (Bodder, 1974).

Special studies on sensitization

Although there are only a few reports on sensitization to tragacanth gum, the available information indicates that tragacanth gum is a powerful allergen capable of causing extremely severe reactions. Allergic reactions may occur as a result of inhalation or oral ingestion (Gelfand, 1943; 1949).

The immunogenicity of tragacanth gum was demonstrated in an in vivo test using a foot pad swelling test in mice. Purification of the gum led to a marked reduction of the immune response (Strobel et al., 1982).

Acute toxicity

The acute oral LD_{50} of 12 food-grade gums (sodium and calcium carragheenate, tragacanth, ghatti, locust bean, arabic, guar, karaya, propylene glycol, alginate, furcellaran, agar agar, and sodium carboxymethyl cellulose) were studied. Each gum was administered by gavage to 5 groups of 10 animals, with 5 males and 5 females in each group. Vehicles utilized were water, mineral oil, corn oil, and soybean oil. The animals were fasted 18 hours prior to dosing with food and water available ad libitum during the 14-day observation period. LD_{50} values observed ranged from 2.6 to 18.0 g/kg, with most values in the 5 to 10 g/kg range. Generally, the rabbit was the most sensitive species and the rat and mouse the least sensitive (Bailey, personal communication to WHO, 1976).

Short-term studies

Rats

Groups of newly-weaned rats (10 per group) were fed a soybean-corn meal diet containing 2% tragacanth gum for 37 days. Tragacanth gum had no effect on the digestibility of the diet, nor was there any significant effect on growth (Vohra et al., 1979).

Tragacanth gum was used in a 6-7 week feeding study to evaluate the effect on adaptive responses of nutritionally-controlled parameters in rats by feeding a fibre-free diet containing increasing additions of polysaccharides (0, 10, 20, and 40%). In general, the supplements reduced growth rates due to lower energy intakes. None of the polysaccharides fed, however, decreased energy utilization. Similarly, all polysaccharides increased small intestine weights by up to about 30% without grossly altering mucosal protein or DNA per unit of length. Concerning the effect on the large intestine, tragacanth gum had a pronounced effect on caecum weight, which increased by factors of 1.8, 2.0, and 4.2 for additions of 10, 20, and 40%, respectively. The degree of the observed changes was determined mainly by the dietary concentration of the polysaccharide and its accessibility to bacterial degradation within the intestinal tract (Elsenhaus et al., 1981).

Groups of 50 male and 50 female Osborne-Mendel rats (approximately 21 days of age) were maintained on diets containing 0, 0.006, 0.06, 0.6, or 6.0% ppm tragacanth gum. After 13 weeks on the test diets, the rats were bred to produce an F_1 generation. The offspring were weaned at day 21 and placed on their respective diets. The animals in the F_0 generation were maintained on the test diets for a total period of 27 weeks. Groups of 50 male and 50 female rats of the F_1 generation were maintained on the test diets for approximately 20 weeks. During the course of the study, body weights and food intake were measured. Reproduction data included the fertility index, total number of progeny, average litter size of pregnant females, total number of liveborn, viability index, survivors to days 4 and 21, weaning index at birth, and weaning weights. At termination of the study, haematological and clinical chemistry studies were carried out. Organ weights were determined, and a complete histological study was made of the principal tissues and organs. Special studies were carried out on liver composition (DNA, RNA, and protein levels), liver DNA synthesis, and intermediary metabolism.

Both males and females in the 6% group showed significantly lower body weights, as well as decreased food efficiency, than the controls. Lower body weights were also observed in the F_1 generation, particularly in the males. Haematological measurements showed no compound-related effects. Only minor effects were noted in the various clinical chemistry parameters. Reproduction data were comparable for test and control animals. Histological studies did not show any compound-related effects. Enlarged livers were noted in the 6% group, but the enlargement was not associated with any significant change in liver composition or with histological changes. The ATP/ADP ratio in liver preparations for F_0 animals was markedly decreased, but this effect was not observed in F_1 animals (Graham et al., 1985).

Chickens
Groups of day-old broiler chickens (7 per group) were fed a soybean-corn diet containing 2% tragacanth gum for 24 days. The dietary intake of 10 chickens was measured daily for the last week of the experimental period. Digestiblity of the test diet was calculated from the dry weights of the feed and excreta. Body weights and the digestibility of the diet were reduced significantly by the inclusion of tragacanth gum in the diet (Vohra et al., 1979).

Quail
Groups of day-old Japanese quail (10 per group) were fed a soybean-corn diet containing 2% tragacanth gum for 36 days. Tragacanth gum did not affect significantly the growth of the quail or the digestibility of the diet (Vohra et al., 1979).

Long-term studies
No information available.

Observations in man
Following a 7-day control period, 5 healthy men ingested 9.9 g tragacanth gum daily (3 x 3.3 g-portion gelled in 200 ml water)

for 32 days. The following measurements were made during the control
period and at the end of the test period: blood glucose, insulin,
serum lipid estimations of cholesterol and HDL cholesterol, phospholip-
ids, triglycerides, haematological indices, and biochemical analyses.
Twenty-four-hour urine samples were collected and tested for sugar,
protein, and blood. Five-day faecal collections were made during days
2-6 of the control period and during days 16-20 of the treatment
period. The tragacanth gum was well-tolerated and no adverse effects
were reported in any of the volunteers. Tragacanth gum had no significant effect on any of the parameters measured with the exception that
intestinal transit time decreased, and faecal wet- and dry-weights were
increased in all subjects at the end of the test period. Four subjects
also showed an increase in faecal fat concentration (Eastwood, et al.,
1984).

Comments

Tragacanth gum is fermented by several strains of bacteria
found in the human colon.

An earlier study showed changes in liver microsomal enzyme
activity and in the oxidative phosphorylation function of heart and
liver mitochondria isolated from rats fed tragacanth gum. However, a
recent study showed no detectable ultrastructural abnormalities in rat
heart or liver and no changes in microsomal protein or P-450 content of
the liver that could be attributable to tragacanth gum.

Tragacanth gum was not mutagenic in bacterial or mammalian
systems. No teratogenic effects were observed in studies in mice,
rats, guinea-pigs, or rabbits. Maternal toxicity observed in rats and
rabbits, at the highest levels tested, may have been due to the mode of
administration (suspended in corn oil), rather than to any innate
toxicity of the gum.

Tragacanth gum fed to rats at dietary levels up to 6% had no
effect on reproductive performance nor on post-partum development of
the pups. The only effects observed in both the F_0 and F_1 generations were lower body weights and decreased feed-efficiency at the 6%
level. Since these effects were not accompanied by any compound-
related histological changes in any tissues or organs, the effects may
be due to the bulking effect of this non-nutritive substance, rather

than any innate toxicity. The highest level tested in this study exceeds the maximum level (5%) recommended for non-nutrients.

Relatively high levels of tragacanth gum were well-tolerated by man. However, possible allergic effects need to be considered.

EVALUATION
Level causing no toxicological effect
Rat: 6000 ppm tragacanth gum in the diet, equivalent to
        3000 mg/kg b.w./day.

Estimate of acceptable daily intake for man
ADI "not specified".

REFERENCES

Anderson, D.M.W., Ashby, P., Busuttil, A., Kempson, S.A., & Lawson, M.E. (1984). Transmission electron microscopy of heart and liver tissues from rats fed with gums arabic and tragacanth. Toxicology Letters, 21, 83-89.

Bachmann, E., Weber, E., Post, M., & Zbinden, G. (1978). Biochemical effects of gum arabic, gum tragacanth and carboxymethyl cellulose in rat heart and liver. Pharmacology, 17, 39-49.

Bodder, R. (1974). Evaluation of chemicals for toxic and teratogenic effects using the chick embryo as a test system. Unpublished report of Warf Institute, Inc. Submitted to WHO by U.S. Food and Drug Administration.

Eastwood, M.A., Brydon, W.G., & Anderson, D.M.W. (1984). The effects of dietary gum tragacanth in man. Toxicology Letters, 21, 73-81.

Elsenhaus, B., Blume, R., & Caspary, W.F. (1981). Long-term feeding of unavailable carboydrate gelling agents. Influence of dietary concentration and microbiological degradation on adaptive responses in the rat. Am. J. Clin. Nutr., 34, 1837-48.

FDRL, 1972.   Teratology evaluation of gum tragacanth in mice, rats,
        hamsters and rabbits.   Unpublished report of Food and Drug
        Research Laboratories, Inc.   Submitted to WHO by U.S. Food
        and Drug Administration.

Frohberg, H., Oettel, H., & Zeller, H. (1969).   Concerning the mech-
        anisms of the fetal toxic effects of tragacanth.   Arch.
        Toxicol., 25, 268-295.

Fujimoto, J.M. (1965).   Effect of gum tragacanth, urethan, and pheno-
        barbital on hexobarbital narcosis in mice.   Toxicol. Appl.
        Pharmacol., 7, 287-290.

Gelfand, H.H. (1943).   The allergenicity of vegetable gums, a case of
        asthma due to tragacanth.   J. Allergy, 14, 203.

Gelfand, H.H. (1949).   Vegetable gums by ingestion in etiology of aller-
        gic disorders.   J. Allergy, 20, 311-321.

Graham, S.L., Friedman, L., & Garthoff, L. (1985).   The subchronic
        effects of gum tragacanth on F_0 and F_1 generation Osborn-
        Mendel rats.   Unpublished report of the Food and Drug Admin-
        istration.   Submitted to WHO by U.S. Food and Drug Adminis-
        tration.

Gohl, B. & Gohl, I. (1977).   The effect of viscous substances on the
        transit time of barley digesta in rats.   J. Sci. Fd. Agric.,
        28, 911-915.

Litton Bionetics (1972).   Summary of mutagenicity screening studies,
        contract FDA71-268, gum tragacanth host-mediated assay,
        cytogenetics, dominant lethal assay, L.B.I. Project No.
        2311.   Unpublished report of Litton Bionetics Inc.
        Submitted to WHO by U.S. Food and Drug Administration.

Litton Bionetics (1977).   Mutagenicity evaluation of gum tragacanth,
        LBI Project No. 20671.   Unpublished report of Litton
        Bionetics Inc. Submitted to WHO by U.S. Food and Drug
        Administration.

Ochuba, G.U. & von Riesen, V.L. (1980). Fermentation of polysaccharides
        by Klebsiellae and other facultative bacilli.   Appl.
        Environ. Microbiol., 39, 988-992.

Riccardi, B.A. & Fahrenbach, M.J. (1965).   Hypocholestrolemic activity
     of mucilaginous polysaccharides in white leghorn cockerels.
     Fed. Proc., 24, 263-265.

Salyers, A.A.,   Vercellotti, J.R.,   West, S.E.H.,   & Wilkins, T.D.
     (1977a).   Fermentation of mucins and plant polysaccharides
     by strains of Bacteroides from the human colon.   Appl.
     Environ. Microbiol., 33, 319-322.

Salyers, A.A.,   Vercellotti, J.R.,   West, S.E.H.,   & Wilkins, T.D.
     (1977b).   Fermentation of mucins and plant polysaccharides
     by anaerobic bacteria from the human colon.   Appl. Environ.
     Microbiol., 34, 529-533.

Strobel, S.,   Ferguson, A.,   & Anderson, D.M.W. (1982).   Immunogenicity
     of food and food additives - In vivo testing of gums arabic,
     karaya and tragacanth.   Toxicology Letters, 14, 247-252.

Vohra, P.,   Shariff, G.,   & Kratzer, F.H. (1979).   Growth inhibitory
     effect of some gums and pectin for Tribolium castaneum
     larvae, chickens, and Japanese quail.   Nutr. Rep. Int., 19,
     463-469.

ANNEXES

REPORTS AND OTHER DOCUMENTS RESULTING FROM MEETINGS OF THE
JOINT FAO/WHO EXPERT COMMITTEE ON FOOD ADDITIVES

1. **General principles governing the use of food additives**
(First report of the Joint FAO/WHO Expert Committee on Food
Additives). FAO Nutrition Meetings Report Series, No. 15,
1958; WHO Technical Report Series, No. 129, 1957 (out of
print).

2. **Procedures for the testing of intentional food additives to
establish their safety for use** (Second report of the Joint
FAO/WHO Expert Committee on Food Additives). FAO Nutrition
Meetings Report Series, No. 17, 1958; WHO Technical Report
Series, No. 144, 1958 (out of print).

3. **Specifications for identity and purity of food additives
(antimicrobial preservatives and antioxidants)** (Third
report of the Joint FAO/WHO Expert Committee on Food
Additives). These specifications were subsequently revised
and published as **Specifications for identity and purity of
food additives**, Vol. I. **Antimicrobial preservatives and
antioxidants**, Rome, Food and Agriculture Organization of
the United Nations, 1962 (out of print).

4.   **Specifications for identity and purity of food additives
          (food colours)** (Fourth report of the Joint FAO/WHO Expert
          Committee on Food Additives). These specifications were
          subsequently revised and published as **Specifications for
          identity and purity of food additives**, Vol. II. **Food
          colours**, Rome, Food and Agriculture Organization of the
          United Nations, 1963 (out of print).

5.   **Evaluation of the carcinogenic hazards of food additives**
          (Fifth report of the Joint FAO/WHO Expert Committee on Food
          Additives). FAO Nutrition Meetings Report Series, No. 29,
          1961; WHO Technical Report Series, No. 220, 1961 (out of
          print).

6.   **Evaluation of the toxicity of a number of antimicrobials
          and antioxidants** (Sixth report of the Joint FAO/WHO Expert
          Committee on Food Additives). FAO Nutrition Meetings Report
          Series, No. 31, 1962; WHO Technical Report Series, No. 228,
          1962 (out of print).

7.   **Specifications for the identity and purity of food
          additives and their toxicological evaluation: emulsifiers,
          stabilizers, bleaching and maturing agents** (Seventh report
          of the Joint FAO/WHO Expert Committee on Food Additives).
          FAO Nutrition Meetings Report Series, no. 35, 1964; WHO
          Technical Report Series, No. 281, 1964 (out of print).

8.   **Specifications for the identity and purity of food addi-
          tives and their toxicological evaluation: food colours and
          some antimicrobials and antioxidants** (Eighth report of the
          Joint FAO/WHO Expert Committee on Food Additives). FAO
          Nutrition Meetings Report Series, No. 38, 1965; WHO
          Technical Report Series, No. 309, 1965 (out of print).

9.   Specifications for identity and purity and toxicological
          evaluation of some antimicrobials and antioxidants. FAO
          Nutrition Meetings Report Series, No. 38A, 1965; WHO/Food/
          Add/24.65 (out of print).

10.   Specifications for identity and purity and toxicological
          evaluation of food colours. FAO Nutrition Meetings Report
          Series, No. 35B, 1966; WHO/Food Add/66.25.

11.   Specifications for the identity and purity of food addi-
          tives and their toxicological evaluation: some anti-
          microbials, antioxidants, emulsifiers, stabilizers, flour-
          treatment agents, acids, and bases (Ninth report of the
          Joint FAO/WHO Expert Committee on Food Additives). FAO
          Nutrition Meetings Report Series, No. 40, 1966; WHO
          Technical Report Series, No. 339, 1966 (out of print).

12.   Toxicological evaluation of some antimicrobials, antioxi-
          dants, emulsifiers, stabilizers, flour-treatment agents,
          acids, and bases. FAO Nutrition Meetings Report Series,
          No. 40A, B, C; WHO/Food Add/67.29.

13.   Specifications for the identity and purity of food addi-
          tives and their toxicological evaluation: some emulsifiers
          and stabilizers and certain other substances (Tenth report
          of the Joint FAO/WHO Expert Committee on Food Additives).
          FAO Nutrition Meetings Report Series, No. 43, 1967; WHO
          Technical Report Series, No. 373, 1967.

14.   Specifications for the identity and purity of food addi-
          tives and their toxicological evaluation: some flavouring
          substances and non-nutritive sweetening agents (Eleventh
          report of the Joint FAO/WHO Expert Committee on Food
          Additives). FAO Nutrition Meetings Report Series, No. 44,
          1968; WHO Technical Report Series, No. 383, 1968.

15.   **Toxicological** evaluation of some flavouring substances and **non-nutritive sweetening agents.** FAO Nutrition Meetings Report Series, No. 44A, 1968; WHO/Food Add/68.33.

16.   **Specifications and criteria for identity and purity of some flavouring substances and non-nutritive sweetening agents.** FAO Nutrition Meetings Report Series, No. 44B, 1969; WHO/Food Add/69.31.

17.   **Specifications for the identity and purity of food additives and their toxicological evaluation: some antibiotics** (Twelfth report of the Joint FAO/WHO Expert Committee on Food Additives). FAO Nutrition Meetings Report Series, No. 45, 1969; WHO Technical Report Series, No. 430 , 1969.

18.   **Specifications for the identity and purity of some antibiotics.** FAO Nutrition Meetings Report Series, No. 45A, 1969; WHO/Food Add/69.34.

19.   **Specifications for the identity and purity of food additives and their toxicological evaluation: some food colours, emulsifiers, stabilizers, anticaking agents, and certain other substances** (Thirteenth report of the Joint FAO/WHO Expert Committee on Food Additives). FAO Nutrition Meetings Report Series, No. 46, 1970; WHO Technical Report Series, No. 445, 1970.

20.   **Toxicological evaluation of some food colours, emulsifiers, stabilizers, anticaking agents, and certain other substances.** FAO Nutrition Meetings Report Series, No. 46A, 1970; WHO/Food Add/70.36.

21.   **Specifications for the identity and purity of some food colours, emulsifiers, stabilizers, anticaking agents, and certain other food additives.** FAO Nutrition Meetings Report Series, No. 46B, 1970; WHO/Food Add/70.37.

22.   **Evaluation of food additives: specifications for the identity and purity of food additives and their toxicological
evaluation: some extraction solvents and certain other
substances; and a review of the technological efficacy of
some antimicrobial agents** (Fourteenth report of the Joint
FAO/WHO Expert Committee on Food Additives). FAO Nutrition
Meetings Report Series, No. 48, 1971; WHO Technical Report
Series, No. 462, 1971.

23.   **Toxicological evaluation of some extraction solvents and
certain other substances.** FAO Nutrition Meetings Report
Series, No. 48A, 1971; WHO/Food Add/70.39.

24.   **Specifications for the identity and purity of some extraction solvents and certain other substances.** FAO Nutrition
Meetings Report Series, No. 48B, 1971; WHO/Food Add/70.40.

25.   **A review of the technological efficacy of some antimicrobial agents.** FAO Nutrition Meetings Report Series, No.
48C, 1971; WHO/Food Add/70.41.

26.   **Evaluation of food additives: some enzymes, modified
starches, and certain other substances: toxicological evaluations and specifications and a review of the technological
efficacy of some antioxidants** (Fifteenth report of the
Joint FAO/WHO Expert Committee on Food Additives). FAO
Nutrition Meetings Report Series, No. 50, 1972; WHO
Technical Report Series, No. 488, 1972.

27.   **Toxicological evaluation of some enzymes, modified
starches, and certain other substances.** FAO Nutrition
Meetings Report Series, No. 50A, 1972; WHO Food Additives
Series, No. 1, 1972.

28.   **Specifications for the identity and purity of some enzymes
        and certain other substances.** FAO Nutrition Meetings
        Report Series, No. 50B, 1972; WHO Food Additives Series, No.
        2, 1972.

29.   **A review of the technological efficacy of some antioxidants
        and synergists.** FAO Nutrition Meetings Report Series, No.
        50C, 1972; WHO Food Additives Series, No. 3, 1972.

30.   **Evaluation of certain food additives and the contaminants
        mercury, lead, and cadmium** (Sixteeth report of the Joint
        FAO/WHO Expert Committee on Food Additives). FAO Nutrition
        Meetings Report Series, No. 51, 1972; WHO Technical Report
        Series, No. 505, 1972, and corrigendum.

31.   **Evaluation of mercury, lead, cadmium, and the food
        additives amaranth, diethylpyrocarbamate, and octyl
        gallate.** FAO Nutrition Meetings Report Series, No. 51A,
        1972; WHO Food Additives Series, No. 4, 1972.

32.   **Toxicological evaluation of certain food additives with a
        review of general principles and of specifications** (Seven-
        teenth report of the Joint FAO/WHO Expert Committee on Food
        Additives). FAO Nutrition Meetings Report Series, No. 53,
        1974; WHO Technical Report Series, No. 539, 1974, and
        corrigendum (out of print).

33.   **Toxicological evaluation of certain food additives includ-
        ing anticaking agents, antimicrobials, antioxidants, emulsi-
        fiers, and thickening agents.** FAO Nutrition Meetings
        Report Series, No. 53A, 1974; WHO Food Additives Series, No.
        5, 1974.

34.   **Specifications for identity and purity of thickening
        agents, anticaking agents, antimicrobials, antioxidants and
        emulsifiers.** FAO Food and Nutrition Paper, No. 4, 1978.

35.   **Evaluation of certain food additives** (Eighteenth report of the Joint FAO/WHO Expert Committee on Food Additives). FAO Nutrition Meetings Report Series, No. 54, 1974; WHO Technical Report Series, No. 557, 1974, and corrigendum.

36.   **Toxicological evaluation of some food colours, enzymes, flavour enhancers, thickening agents, and certain other food additives.** FAO Nutrition Meetings Report Series, No. 54A, 1975; WHO Food Additives Series, No. 6, 1975.

37.   **Specifications for the identity and purity of some food colours, flavour enhancers, thickening agents, and certain food additives.** FAO Nutrition Meetings Report Series, No. 54B, 1975; WHO Food Additives Series, No. 7, 1975.

38.   **Evaluation of certain food additives: some food colours, thickening agents, smoke codensates, and certain other substances** (Nineteenth report of the Joint FAO/WHO Expert Committee on Food Additives). FAO Nutrition Meetings Report Series, No. 55, 1975; WHO Technical Report Series, No. 576, 1975.

39.   **Toxicological evaluation of some food colours, thickening agents, and certain other substances.** FAO Nutrition Meetings Report Series, No. 55A, 1975; WHO Food Additives Series, No. 8, 1975.

40.   **Specifications for the identity and purity of certain food additives.** FAO Nutrition Meetings Report Series, No. 55B, 1976; WHO Food Additives Series, No. 9, 1976.

41.   **Evaluation of certain food additives** (Twentieth report of the Joint FAO/WHO Expert Committee on Food Additives). FAO Food and Nutrition Series, No. 1, 1976; WHO Technical Report Series, No. 599, 1976.

42.   **Toxicological evaluation of certain food additives.** WHO
           Food Additives Series, No. 10, 1976.

43.   **Specifications for the identity and purity of some food
           additives.** FAO Food and Nutrition Series, No. 1B, 1977;
           WHO Food Additives Series, No. 11, 1977.

44.   **Evaluation of certain food additives** (Twenty-first report
           of the Joint FAO/WHO Expert Committee on Food Additives).
           WHO Technical Report Series, No. 617, 1978.

45.   **Summary of toxicological data of certain food additives.**
           WHO Food Additives Series, No. 12, 1977.

46.   **Specifications for identity and purity of some food addi-
           tives, including antioxidants, food colours, thickeners, and
           others.** FAO Nutrition Meetings Report Series, No. 57, 1977.

47.   **Evaluation of certain food additives and contaminants**
           (Twenty-second report of the Joint FAO/WHO Expert Committee
           on Food Additives). WHO Technical Report Series, No. 631,
           1978.

48.   **Summary of toxicological data of certain food additives and
           contaminants.** WHO Food Additives Series, No. 13, 1978.

49.   **Specifications for the identity and purity of certain food
           additives.** FAO Food and Nutrition Paper, No. 7, 1978.

50.   **Evaluation of certain food additives** (Twenty-third report
           of the Joint FAO/WHO Expert Committee on Food Additives).
           WHO Technical Report Series, No. 648, 1980, and corrigenda.

51.   **Toxicological evaluation of certain food additives.** WHO
           Food Additives Series, No. 14, 1980.

52.   Specifications for identity and purity of food colours, flavouring agents, and other food additives. FAO Food and Nutrition Paper, No. 12, 1979.

53.   Evaluation of certain food additives (Twenty-fourth report of the Joint FAO/WHO Expert Committee on Food Additives). WHO Technical Report Series, No. 653, 1980.

54.   Toxicological evaluation of certain food additives. WHO Food Additives Series, No. 15, 1980.

55.   Specifications for identity and purity of food additives (sweetening agents, emulsifying agents, and other food additives). FAO Food and Nutrition Paper, No. 17, 1980.

56.   Evaluation of certain food additives (Twenty-fifth report of the Joint FAO/WHO Expert Committee on Food Additives). WHO Technical Report Series, No. 669, 1981.

57.   Toxicological evaluation of certain food additives. WHO Food Additives Series, No. 16, 1981.

58.   Specifications for identity and purity of food additives (carrier solvents, emulsifiers and stabilizers, enzyme preparations, flavouring agents, food colours, sweetening agents, and other food additives). FAO Food and Nutrition Paper, No. 19, 1981.

59.   Evaluation of certain food additives and contaminants (Twenty-sixth report of the Joint FAO/WHO Expert Committee on Food Additives). WHO Technical Report Series, No. 683, 1982.

60.   Toxicological evaluation of certain food additives. WHO Food Additives Series, No. 17, 1982.

61.   **Specifications for the identity and purity of certain food
          additives.** FAO Food and Nutrition paper, No. 25, 1982.

62.   **Evaluation of certain food additives and contaminants**
          (Twenty-seventh report of the Joint FAO/WHO Expert Committee
          on Food Additives). WHO Technical Report Series, No. 696,
          1983, and corrigenda.

63.   **Toxicological evaluation of certain food additives and
          contaminants.** WHO Food Additives Series, no. 18, 1983.

64.   **Specifications for the identity and purity of certain food
          additives.** FAO Food and Nutrition Paper, No. 28, 1983.

65.   **Guide to specifications--General notices, general methods,
          identification tests, test solutions, and other reference
          materials.** FAO Food and Nutrition Paper, No. 5, Rev. 1,
          1983.

66.   **Evaluation of certain food additives and contaminants**
          (Twenty-eighth report of the Joint FAO/WHO Expert Committee
          on Food Additives). WHO Technical Report Series, No. 710,
          1984.

67.   **Toxicological evaluation of certain food additives and
          contaminants.** WHO Food Additives Series, No. 19, 1984.

68.   **Specifications for the identity and purity of certain food
          additives.** FAO Food and Nutrition paper, No. 31/1, 1984.

69.   **Specifications for the identity and purity of certain food
          additives.** FAO Food and Nutrition Paper, No. 31/2, 1984.

70.   **Evaluation of certain food additives and contaminants**
          (Twenty-ninth report of the Joint FAO/WHO Expert Committee
          on Food Additives). WHO Technical Report Series, No. 733,
          1986.

ABBREVIATIONS USED IN THE MONOGRAPHS

ADI	acceptable daily intake
BUN	blood urea nitrogen
b.w.	body weight
CD_{50}	convulsive dose, median
CHO	Chinese hampster ovary
DCI	immobilized glucose isomerase from S. rubiginosis
FAO	Food and Agriculture Organization of the United Nations
FFA	free fatty acids
g	gram
GOT	see SGOT
GPM	alpha-D-glucopyranosido-1,6-mannitol
GPS	alpha-D-glucopyranosido-1,6-sorbitol
GPT	see SGPT
Hb	haemoglobin
HFE	non-immobilized glucose isomerase from S. rubiginosis
HGS	hydrogenated glucose syrups
IARC	International Agency for Research on Cancer
i.m.	intramuscular
i.p.	intraperitoneal
IPCS	International Programme on Chemical Safety

IRI	immunoreactive insulin
i.v.	intravenous
JECFA	Joint FAO/WHO Expert Committee on Food Additives
kg	kilogram
LD_{50}	lethal dose, median
LDH	lactate dehydrogenase
4-MEI	4-methylimidazole
μg	microgram
μl	microlitre
μM	micromolar
mg	milligram
ml	millilitre
mM	millimolar
MTD	maximum tolerated dose
NAG	N-acetylglucosaminidase
nM	nanomolar
NOEL	no observed effect level
OCT	ornithine carbamoyl transferase
PCV	haematocrit
per os	by mouth
RBC	red blood cell (erythrocyte count)
SAP	serum alkaline phosphatase
s.c.	subcutaneous
SG	specific gravity
SGOT	serum glutamate-oxaloacetate transaminase
SGPT	serum glutamate-pyruvate transaminase
T_3	triiodothyronine
T_4	thyroxine
TG	triglycerides
THI	2-acetyl-4(5)-tetrahydroxybutylimidazole
WBC	white blood cell (total leukocyte count)
WHO	World Health Organization
w/w	weight/weight

JOINT FAO/WHO EXPERT COMMITTEE ON FOOD ADDITIVES
Geneva, 3-12 June 1985

Members invited by FAO

Dr I. Chakravarty, Professor and Head, Department of Biochemistry and
Nutrition, All India Institute of Hygiene and Public Health,
Calcutta, India

Dr W.H.B. Denner, Head, Food Composition and Information Unit,
Food
Sciences Division, Ministry of Agriculture, Fisheries and
Food, London, England (Vice-Chairman)

Dr S.W. Gunner, Director General, Food Directorate, Health Protection
Branch, Health and Welfare Canada, Ottawa, Canada

Professor K. Kojima, College of Environmental Health, Azabu University, Sagamihara-Shi, Kanagawa-Ken, Japan

Dr W. Kroenert, Head, Food Chemistry Division, Max von Pettenkofer
Institute, Federal Office of Public Health, Berlin (West)

Dr R. Mathews, Director, Food Chemicals Codex, National Academy of
Sciences, Washington, DC, USA

Dr J.P.M. Modderman, Food Additives Chemistry Evaluation Branch, Center
for Food Safety and Applied Nutrition, Food and Drug Administration, Washington, DC, USA

Professor F.J. Pellerin, Faculty of Pharmacy, Université de Paris XI,
Centre hospitalier Corentin-Celton,   Issy-les-Moulineaux,
France

Members invited by WHO

Dr H. Blumenthal, Director, Division of Toxicology, Center for Food
Safety and Applied Nutrition, Food and Drug Administration,
Washington, DC, USA

Dr A.H. El-Sebae, Chairman, Pesticides Division, Faculty of Agriculture, Alexandria University, Alexandria, Egypt

Professor P.E. Fournier, Professor of Clinical Toxicology, Hôpital
Fernand Widal, Paris, France

Dr B. MacGibbon, Senior Principal Medical Officer, Division of Toxicology and Environmental Protection, Department of Health and
Social Security, London, England

Professor K.A. Odusote, Associate Professor of Medicine, College of
Medicine, University of Lagos, Lagos, Nigeria (Rapporteur)

Dr P. Pothisiri, Director, Food Control Division, Food and Drug Administration, Ministry of Public Health, Bangkok, Thailand

Professor M.J. Rand, Head, Department of Pharmacology, University of
Melbourne, Victoria, Australia (Chairman)

*Dr V.A. Tutelyan, Deputy Director, Institute of Nutrition, Academy of
Medical Sciences of the USSR, Moscow, USSR

---

* Invited but unable to attend

Secretariat

Dr J.R.P. Cabral, Scientist, Unit of Mechanisms of Carcinogenesis, International Agency for Research on Cancer, Lyons, France

Mr A. Feberwee, Chairman, Codex Committee on Food Additives; and Deputy Director, Nutrition and Quality Affairs, Ministry of Agriculture and Fisheries, The Hague, The Netherlands (Member of FAO Secretariat)

Professor C.L. Galli, Professor of Experimental Toxicology, Institute of Pharmacology and Pharmacognosy, University of Milan, Milan, Italy (WHO Temporary Adviser)

Dr W. Grunow, Head, Division of Food Toxicology, Max von Pettenkofer Institute, Federal Office of Public Health, Berlin (West) (WHO Temporary Adviser)

Mr R. Haigh, Principal Administrator, Commission of the European Communities, Brussels, Belgium (WHO Temporary Adviser)

Dr Y. Hayashi, Director, Division of Pathology, Biological Safety Research Centre, National Institute of Hygienic Sciences, Tokyo, Japan (WHO Temporary Adviser)

Dr J.L. Herrman, Division of Food and Color Additives, Center for Food Safety and Applied Nutrition, Food and Drug Administration, Washington, DC, USA (WHO Consultant)

Dr F. Käferstein, Responsible Officer, Food Safety Programme, Division of Environmental Health, WHO, Geneva, Switzerland

Dr N. Rao Maturu, Food Standards Officer, Joint FAO/WHO Food Standards Programme, FAO, Rome, Italy

Dr M. Mercier, Manager, International Programme on Chemical Safety, Division of Environmental Health, WHO, Geneva, Switzerland

Dr A.W. Randell, Nutrition Officer (Food Science), Food Policy and Nutrition Division, FAO, Rome, Italy (FAO Joint Secretary)

Dr S.I. Shibko, Associate Director for Regulatory Evaluation, Division of Toxicology, Center for Food Safety and Applied Nutrition, Food and Drug Administration, Washington, DC, USA (WHO Temporary Adviser)

Dr P. Shubik, Senior Research Fellow, Green College, Oxford, England (WHO Temporary Adviser)

Professor R. Truhaut, Professor Emeritus of Toxicology, Department of Toxicology and Industrial Hygiene, René Descartes University, Paris, France (WHO Temporary Adviser)

Dr G. Vettorazzi, Senior Toxicologist, International Programme on Chemical Safety, Division of Environmental Health, WHO, Geneva, Switzerland (WHO Joint Secretary)

Dr R. Walker, Department of Biochemistry, University of Surrey, Guildford, Surrey, England (WHO Temporary Adviser)

ACCEPTABLE DAILY INTAKES, OTHER TOXICOLOGICAL
RECOMMENDATIONS AND INFORMATION ON SPECIFICATIONS

	Specifications [1]	ADI for man and other toxicological recommendations

A.   **Specific food additives**

Enzyme preparations and enzyme immobilizing agents

Carbohydrase (α-amylase) from Bacillus licheniformis	R	ADI not specified [2]
Glucose isomerase (immobilized) from Actinoplanes missouriensis	S	Acceptable [3]
Glucose isomerase from Bacillus coagulans	S	No ADI allocated [4]
Glucose isomerase (immobilized) from Bacillus coagulans	S	Acceptable [3]
Glucose isomerase (immobilized) from Streptomyces olivaceus	S	Acceptable [3]
Glucose isomerase (immobilized) from Streptomyces olivochromogenes	S	Acceptable [3]
Glucose isomerase from Streptomyces rubiginosus	R	No ADI allocated [4]
Glucose isomerase (immobilized) from Streptomyces rubiginosus)	R	Acceptable [3]
Polyethylenimine	–	Suitable [5]

Flavouring agent
Benzyl acetate	S	0–5 [6] mg/kg b.w.

	Specifications[1]	ADI for man and other toxicological recommendations
Flour treatment agent		
Chlorine	S	2.5g Cl_2/kg flour[7]
Food acids and their salts		
Aluminium ammonium sulfate	N	0–0.6 mg/kg b.w.[6],[8]
Aluminium, calcium, magnesium, potassium, and sodium salts of capric, caprylic, lauric and oleic acids	O	No ADI allocated[9]
Ammonium succinate	O	No ADI allocated[9]
Calcium adipate	O	No ADI allocated[9]
Calcium aluminium silicate (previously aluminium calcium silicate)	S	ADI "not specified"[2],[10]
Calcium fumarate	O	No ADI allocated[9]
Calcium hydrogen carbonate	O	No ADI allocated[9]
Calcium succinate	O	No ADI allocated[9]
Dipotassium guanylate	N	ADI "not specified"[2],[12]
Dipotassium inosinate	N	ADI "not specified"[2],[13]
Ferric ammonium citrate	S	0–0.8 mg/kg b.w.[11]
Guanylic acid	N	ADI "not specified"[2],[12]
Inosinic acid	N	ADI "not specified"[2],[13]
Magnesium acetate	O	No ADI allocated[9]
Magnesium adipate	O	No ADI allocated[9]
Magnesium citrate	O	No ADI allocated[9]
Magnesium succinate	O	No ADI allocated[9]
Monomagnesium phosphate	O	No ADI allocated[9]
Potassium aluminosilicate	O	No ADI allocated[9]
Potassium fumarate	O	No ADI allocated[9]
Potassium succinate	O	No ADI allocated[9]
Potassium sulfate	N	ADI "not specified"[2]
Potassium sulfite	N	0–0.7 mg/kg b.w.[14]
Sodium aluminium polyphosphate	O	No ADI allocated[9]
Sodium sorbate	O	0–25 mg/kg b.w.[15]
Food colours		
Brown FK	N	0–0.075 mg/kg b.w.[6]
Caramel colours		
Class I	R,T	ADI "not specified"[2]
Class II	N,T	No ADI allocated[16]
Class III	R,T	0–200 mg/kg b.w. (0–150 mg/kg b.w. on solids basis)

	Specifications [1]	ADI for man and other toxicological recommendations
Class IV	R,T	0-200 mg/kg b.w. (0-150 mg/kg b.w. on solids basis)
Carthamus yellow	R,T	No ADI allocated [16]
Fast green FCF	R	0-12.5 mg/kg b.w. [6]
Saffron	R,T	Food ingredient [17]
Sweetening agents		
Hydrogenated glucose syrups	R	ADI "not specified" [2]
Isomalt	R	ADI "not specified" [2]
Mannitol	T	0-50 mg/kg b.w. [6]
Thaumatin	S	ADI "not specified" [2]
Thickening agents		
Dammer gum	S,T	No ADI allocated [16]
Ethylhydroxyethyl cellulose	R,T	0-25 mg/kg b.w. [6,18]
Gum ghatti	R	No ADI allocated [16]
Karaya gum	R	0-20 mg/kg b.w. [6]
Oat gum	0	No ADI allocated [16]
Tragacanth gum	R	ADI "not specified" [2]
Xanthan gum	S	0-10 mg/kg b.w.
Miscellaneous food additives		
Bone phosphate	R	70 mg/kg b.w. [19]
Carbon dioxide	R,T	ADI "not specified" [2]
Nitrous oxide	R	Acceptable [20]
Polyvinylpyrrolidone (PVP)	R,T	0-25 mg/kg b.w. [6]
Quillaia extract	N,T	0-5 mg/kg b.w.
Sodium thiocyanate S		Decision postponed

B. **Contaminants**

Ethylenimine   –   Provisional acceptance [21]

Specifications only [1]

Acesulfame potassium	S
Ammonium hydrogen carbonate	R
Ammonium polyphosphate	S,T
Butylated hydroxyanisole	R
Calcium polyphosphates	S
Calcium, potassium, and sodium salts of myristic, palmitic, and stearic acids	R

Specifications only[1]

Carrageenan	R
Diethyleneglycol monomethylether	S,T
Diethyl tartrate	R
Ethyl alcohol (previously ethanol)	R
Ethylhydroxymethyl cellulose	R
Eugenyl methyl ether	R
Gum arabic	R
Hydrogen peroxide	R
Hydroxypropyl cellulose	R
Hydroxypropylmethyl cellulose	R
Insoluble polyvinylpyrrolidone	R
Methylethyl cellulose	S
Modified starches:	
Acetylated distarch adipate	R
Acetylated distarch phosphate	R,T
Hydroxypropyl distarch phosphate	R
Pentapotassium triphosphate	R
Polydimethylsiloxane	R
Saccharin	R
Sodium aluminium phosphate, acidic	R
Sorbitan monolaurate	R,T
Sorbitan monooleate	R,T
Sorbitol	R
Sucrose acetate isobutyrate	R,T
Turmeric oleoresin	N

Notes to Annex 4

1.   N, new specifications prepared; O, specifications not prepared; R, existing specifications revised; S, specifications exist, revision not considered or not required; and T, the existing, new or revised specifications are tentative and comments are invited.

2.   ADI "not specified" means that, on the basis of the available data (chemical, biochemical, toxicological, and other), the total daily intake of the substance, arising from its use at the levels necessary to achieve the desired effect and from its acceptable background in food, does not, in the opinion of the Committee, represent a hazard to health. For that reason, and for the reasons stated in the individual evaluations, the establishment of an acceptable daily intake (ADI) expressed in numerical form is not deemed necessary.

3.   Acceptable for use in food processing.

4.   No information was available on the food use of this enzyme.

5.   Polyethylenimine is considered to be a suitable substance for use as an immobilizing agent in the production of immobilized enzymes. (See also note 21.)

6.   Temporary acceptance.

7.   Acceptable level for treatment of flours for cake manufacturing.

8.   Group ADI for aluminium salts expressed as aluminium.

9.   No information was available on the food use of these salts.

10.   Group ADI for silicon dioxide and certain silicates (aluminium, calcium, and sodium aluminosilicate); the previous ADI "not limited" was changed to ADI "not specified".

11.   Included in the group maximum tolerable daily intake for iron.

12.   Group ADI for guanylic acid and its calcium, dipotassium, and disodium salts.

13.   Group ADI for inosinic acid and its calcium, dipotassium, and disodium salts.

14.   Group ADI for sulfur dioxide and sulfites (sodium and potassium metabisulfites, sodium sulfite, and sodium hydrogen sulfite expressed as sulfur dioxide).

15.   Group ADI for sorbic acid and its calcium, potassium, and sodium salts expressed as sorbic acid.

16.   Insufficient information available on its toxicology and chemical composition.

17.   This substance is regarded as a food rather than as a food additive.

18.   Group ADI for modified celluloses.

19.   This figure represents the maximum tolerable daily intake (MTDI) of phosphates expressed as phosphorus; it applies to the sum of phosphates naturally present in food and the additives listed in Annex 4 of the twenty-sixth report (WHO Technical Series, No. 683, 1982). It also applies to diets that are nutritionally adequate in respect of calcium. However, if the calcium intake were high, the intake of phosphate could be proportionately higher; the reverse relationship would also apply.

20.   The food use of nitrous oxide as a propellant is acceptable.

21.   Acceptable on condition that the amount of ethylenimine
           migrating into food is reduced to the lowest technically
           attainable level.